ORIGINS

OF

21ST-CENTURY
SPACE TRAVEL

ORIGINS

OF

21ST-CENTURY SPACE TRAVEL

*A History of NASA's Decadal
Planning Team and the Vision for
Space Exploration,
1999–2004*

GLEN R. ASNER
STEPHEN J. GARBER

National Aeronautics and Space Administration

Office of Communications
NASA History Division
Washington, DC 20546

NASA SP-2019-4415

Library of Congress Cataloging-in-Publication Data

Names: Asner, Glen R., 1970– | Garber, Stephen J.
Title: Origins of 21st-century space travel: a history of NASA's Decadal Planning Team and the Vision for Space Exploration, 1999–2004 / by Glen R. Asner and Stephen J. Garber.
Description: Washington, DC : National Aeronautics and Space Administration, Office of Communications, NASA History Division, [2018] | Series: NASA history series | Series: NASA SP ; 2019-4415
Identifiers: LCCN 2015008003
Subjects: LCSH: Astronautics and state—United States—History—21st century. | Astronautics—United States—History—21st century. | Outer space—Exploration—Planning—History—21st century. | United States. National Aeronautics and Space Administration. Decadal Planning Team.
Classification: LCC TL789.8.U5 A77 2018 | DDC 629.40973/090511—dc23
LC record available at *https://lccn.loc.gov/2015008003*

ABOUT THE COVER: The background image depicting astronomical phenomena is from iStockPhoto. The image of a spacecraft is an artist's conception of the Orion capsule, designed to carry humans into deep space, which evolved from the 2004 Vision for Space Exploration.

This publication is available as a free download at
http://www.nasa.gov/ebooks.

ISBN 978-1-62683-045-5

CONTENTS

FOREWORD

William H. Gerstenmaier
NASA ASSOCIATE ADMINISTRATOR
HUMAN EXPLORATION AND OPERATIONS MISSION DIRECTORATE

"Plans are worthless, but planning is everything."
—*Dwight D. Eisenhower, 1957*

LONG-TERM PLANNING is difficult for any organization, but a government organization has a special set of challenges it must deal with. Congress, which appropriates funding and dictates priorities for executive branch agencies, operates on a one-year financial cycle and a two-year political rhythm. The President, to whom all federal executive agencies are directly accountable, changes administrations once every four years, even when the President is reelected. It is against this backdrop that the National Aeronautics and Space Administration (NASA) seeks to develop systems that may take 1, 2, or even 10 or 20 years to develop. The technologies of these systems are independent of political whims, but their engineers, program managers, and agency leaders are not. Even the five-year budget planning horizon favored by government analysts is merely half that of the National Academies' Decadal Survey, which is a key input for any NASA science mission.

When developing a mission, it is NASA's challenge to manage not only the immutable laws of physics, but also the (at times) impenetrable laws of politics that are equally important. The scale of a flagship-class science mission or a human exploration mission simply is not commensurate with the way most government agencies are organized to operate. And yet, for nearly 60 years, NASA has continued to expand the presence of humans in the solar system, both robotically and in person, to become more capable in space than ever before.

One reason for NASA's success has been the hard-won and hard-earned lessons of the past. Experience time and again proved that even the best-laid plans, whose technical and engineering rationale were beyond reproach, could become increasingly vulnerable to charges of inefficiencies and irrelevancies over time. In the realm of human spaceflight, whose ultimate horizon goal is to land humans on the surface of Mars, NASA has embraced a more flexible approach than ever before, seeking to execute what is possible in the near term, plan for what is likely in the midterm, and aim for what is desired in the long term. In this way, the program can move forward through technical and political challenges without losing its core mission.

Glen R. Asner and Stephen J. Garber's new volume is a timely, well-documented, and provocative account of exactly how NASA has begun embracing flexibility. The book looks at one series of mission planning efforts, in a specific technical and political context, starting with the NASA Decadal Planning Team (DPT) and culminating in the 2004 Vision for Space Exploration (VSE). Asner and Garber put NASA's past mission planning efforts into context, looking at both the effect of a change in administration on mission planning and the poignant and tragic effects of replanning after the Columbia accident in 2003. This book focuses on the period between 1999 and 2004, showing how the 2004 VSE was substantially shaped by internal NASA planning efforts, and is in many ways a cautionary tale reminding current managers and future leaders not to become complacent in their planning assumptions.

Asner and Garber's book also reflects on what NASA has accomplished through building the International Space Station (ISS) with the help of the Space Shuttle. The planning efforts described in the book were significantly shaped by the tragedy of the loss of Columbia, which created uncertainties about how long it would take to build the ISS using the Shuttle and how that work would affect broader exploration plans. The completion of the ISS and the retirement of the Space Shuttle enabled NASA to begin the development of the Space Launch System and Orion crew capsule and to shift the focus of work on the ISS toward enabling future exploration efforts.

There are insights throughout the book that will be of value to technical and policy practitioners as well as those with a serious interest in space. Those who dream of affecting national policy need to know the dense policy and organizational thickets they will encounter as surely as they will struggle against gravity. By giving a thorough, yet accessible work of historical scholarship, I hope the authors will help all of us, including the next generation of engineers and mission planners, to sustain the progress we've made.

1
INTRODUCTION

ON 14 JANUARY 2004, a few weeks shy of the one-year anniversary of the Space Shuttle Columbia accident, President George W. Bush stood at a podium in the NASA Headquarters auditorium in Washington, DC, to deliver a major announcement. After thanking several luminaries in attendance, the President spoke of the role of exploration in American history, the Earthly benefits of space investments, and the importance of current programs, from the Space Shuttle and International Space Station (ISS) to the scientific exploration of planets and the universe. As a segue to his announcement, he lamented the failure of the civilian space program over the past 30 years to develop a new vehicle to replace the Shuttle and to send astronauts beyond low-Earth orbit (LEO). The President then revealed the purpose of his visit, declaring that the mission of NASA henceforth would be "to explore space and extend a human presence across our solar system." Although the plan was an explicit expansion of NASA's publicly stated long-term goals, President Bush suggested that it would cause few disruptions since it would rely on "existing programs and personnel" and would be implemented in a methodical fashion, with the Agency progressing at a steady pace toward each major milestone.[1]

The initial, near-term goals of what became known as the Vision for Space Exploration (VSE) were straightforward. The President requested that NASA complete the assembly of the Space Station by 2010 and then use it as a

---

1. President George W. Bush, *Vision for Space Exploration* (speech, NASA Headquarters, Washington, DC, 14 January 2004). See *http://history.nasa.gov/Bush%20SEP.htm* for a transcript of this speech.

President George W. Bush announces his administration's Vision for Space Exploration policy in the NASA Headquarters auditorium on 14 January 2004. (NASA 20040114_potus_07-NASAversion)

laboratory for investigating the impact of space on the human body. Under the plan, NASA would return the Shuttle to flight and keep it flying only until the completion of the station in 2010. The President's second set of goals hinged on the development of a new space vehicle, the Crew Exploration Vehicle (CEV), to ferry astronauts to the ISS, the Moon, and ultimately to other planets. The President set 2008 as the date for the first test of the CEV and no later than 2014 as the date for the first CEV flight with a human occupant. Returning humans to the surface of the Moon with the CEV by 2020, but possibly as early as 2015, was the third and last of Bush's major near-term goals for the space program. Allowing NASA as long as 16 years to reach the Moon, the plan could be seen as lacking ambition and a sense of urgency, but critics could not accuse Bush of failing to articulate clear goals for the early phases of his plan.

The second half of the President's speech was the mirror image of the first half in the sense that it lacked detailed timetables and milestones, but it did not lack ambition. Extended operations on the Moon, Bush revealed, would serve as "a logical step," a proving ground, for "human missions to Mars and to worlds beyond."[2] The President expressed optimism that lunar surface operations involving human explorers would provide opportunities for NASA to develop new capabilities, such as assembling spacecraft on the Moon and utilizing lunar

---

2. Bush, *Vision* speech.

resources, which might dramatically reduce the cost of spaceflight to other locations in the solar system.

Without committing to deadlines or destinations, Bush explained the general approach he expected the VSE to follow. Robotic spacecraft, including orbiters, probes, and landers, would serve as the "advanced [sic] guard to the unknown," investigating the conditions of the solar system and other planets and returning remote sensing data to Earth for analysis. With a better understanding of the conditions they would encounter, human explorers would follow the robotic "trailblazers" to more distant locations.[3] In emphasizing that he was setting in motion a "journey, not a race," the President implicitly contrasted his plan with John F. Kennedy's Apollo Moon shot announcement more than 40 years earlier, as well as President George H. W. Bush's Space Exploration Initiative (SEI), an ambitious plan for reaching Mars that NASA abandoned in the wake of revelations about the projected cost of the plan.[4] The "journey, not a race" theme also anticipated questions that might emerge regarding international participation and the seemingly meager budget increase, just one billion dollars spread over five years, that the President promised to provide NASA to begin implementing the VSE.

Fact sheets, summary documents, and talking points compiled to support the President's announcement elaborated on the motivations and goals for the VSE. The main White House summary document, titled "A Renewed Spirit of Discovery," explained that the President sought to advance multiple national interests—science, economic growth, and national security—with his plan for reinvigorating the civil space program. To serve these broad interests, the President directed NASA to pursue "a sustained and affordable human and robotic program to explore the solar system and beyond."[5] As the President noted in his speech, NASA would operate on the surface of the Moon, developing

---

3. Ibid.

4. For more on SEI, see *http://history.nasa.gov/sei.htm*. Thor Hogan's *Mars Wars: The Rise and Fall of the Space Exploration Initiative* (NASA SP-2007-4410) analyzes the failure of SEI. The explicit "lessons learned" about SEI come from p. 3 of Vision Roll Out Action Item List, Vision-General folder, Joe Wood files, NASA Historical Reference Collection, NASA History Division, NASA Headquarters, Washington, DC 20546 (hereafter referred to as the NASA HRC). Not surprisingly, the Clinton administration also was reluctant to have another SEI-style political fiasco. See, for example, Alan Ladwig's personal notes, 24 March 1999, NASA HRC. Mark Albrecht, *Falling Back to Earth: A First Hand Account of the Great Space Race and the End of the Cold War* (New Media Books: place of publication unknown, 2011) is a memoir that covers Albrecht's time at the National Space Council, including SEI.

5. President George W. Bush, "A Renewed Spirit of Discovery: The President's Vision for U.S. Space Exploration," January 2004, available at *http://history.nasa.gov/renewedspiritofdiscovery.pdf*. For the quotation, see p. 5.

new technologies, knowledge, and infrastructure, as a step toward a more ambitious plan to send humans to Mars and other destinations.

NASA talking points went into greater detail on the motivations and goals for the VSE and used some phrases that the White House did not include in the President's speech or supporting documents.[6] A brief, three-page set of talking points, for example, placed greater emphasis on the scientific possibilities of the VSE, explaining that human and robotic explorers would continue NASA's quest to "answer ageless questions on the origins of the universe and the possibility of life elsewhere" and that NASA would "not forsake its important work in improving the nation's aviation system, in education, in Earth science, and in fundamental space science."[7] Such talking points were at least partly directed toward scientists who were concerned about the lack of any serious discussion of the role of science in the President's speech. The NASA talking points also explained the role of the Moon and the overall exploration plan using terminology familiar to space exploration advocates:

> We will use the Moon as a stepping-stone to enable sustained future human and robotic exploration of Mars and other destinations. The lunar activities will further science and will develop and test new approaches, technologies, and systems, including use of lunar and other space resources, to support sustained human exploration.[8]

To the general public and many at NASA who listened to the speech or read the supporting documents, the details of the plan mattered less than the mere fact that the President had announced an ambitious exploration program that included returning to the Moon and eventually sending humans to Mars. Yet the details—the role of the Moon and Mars; the path toward other destinations; timetables; costs; the mix of robotics and humans; and the scientific, economic, and technological goals of the program—all mattered a great deal, both for determining whether the plan would withstand early criticism and whether it could be sustained in the future, through good times and bad.

The purpose of this study is to trace the ideas, events, and policy debates at NASA and the White House that informed the choices the Bush administration made for the VSE and the future of human exploration at NASA. Although journalists have written numerous articles, there is a surprising

---

6. "Vision talking points," 14 January 2004; Responses to Questions (RTQ) 04-006, 14 January 2004; and RTQ 04-005, 14 January 2004, NASA HRC.

7. "Vision talking points," 14 January 2004.

8. Ibid.

absence of historical works on the VSE that document their primary sources.[9] Considerable gaps remain in our understanding of why the VSE, as President Bush announced it on 14 January 2004, developed as it did. Why, for example, was the Moon so prominent in the VSE? On what basis did the White House reach the decision that supporting an extended presence on the Moon was a more effective and feasible approach to sending humans to other planets than a direct flight? What other options for destinations and timetables did the administration consider? To what extent did the idea of an exploration plan involving "humans and robots" differ from other exploration plans? Why did the administration identify science, economics, and national security as the motivations for the VSE? What priority did the White House place on each of these motivations? Why did the administration provide what many at NASA considered a meager budget—$11 billion reprogrammed from other major NASA programs and $1 billion in additional funds—for the first five years of the VSE?

Along these lines, the role of NASA in the policy development process is not well understood, nor is the difference between the policy President Bush announced and the exploration concepts NASA promoted prior to the Columbia accident and in policy debates that led directly to the VSE. The VSE emerged from a formal, deliberative process in which senior White House advisors, cabinet officials, and staff members of the Executive Office of the President weighed NASA plans for exploration against alternatives conceived by the Office of Management and Budget (OMB), the Council of Economic Advisors (CEA), and the Office of Science and Technology Policy (OSTP). Staff members of the National Security Council (NSC) managed the policy process and mediated disputes between NASA and representatives of other agencies and offices over the details of the policy during its development.

The competition of ideas and White House resistance to changes that would raise NASA's budget significantly meant that the ultimate policy represented a compromise between the interests involved. Yet NASA came to the discussions with clear ideas on how to go about formulating a new long-term space strategy that included both human and robotic space exploration. Before discussions began over the VSE, NASA officials had well-formulated plans that included prioritized goals, mission architectures, approaches, justifications, technological options, and even public-relations concepts. This was not the first time that NASA anticipated White House plans for space—NASA engineers

---

9. Frank Sietzen, Jr., and Keith L. Cowing wrote *New Moon Rising: The Making of America's New Space Vision and the Remaking of NASA* (Burlington, Ontario: Apogee Books, 2004). This journalistic account of the VSE was quickly assembled in 2004 and cites no sources, thus rendering many of its descriptions of events impossible to verify.

had considered various ways to win the space race in the 1960s before President Kennedy publicly announced his bold plans for Apollo in May 1961—but in the case discussed here, NASA sponsored a study team specifically in anticipation of a future opportunity, what some political scientists term a "policy window."[10]

NASA's preparedness was due in part to an embargoed human exploration planning effort that had begun several years before the Columbia accident. The Decadal Planning Team (DPT), which began in 1999 under NASA Administrator Daniel Goldin, and its successor planning group, the NASA Exploration Team (NEXT), laid the groundwork for NASA's participation in the VSE development process. The experience greatly raised the level of knowledge of all aspects of exploration planning of a cadre of NASA scientists and engineers at Headquarters and elsewhere who later played critical roles in the VSE development process. Off-the-shelf DPT and NEXT plans, moreover, largely formed the basis of NASA proposals put before the White House.

Both DPT and the VSE were focused on creating a new paradigm for spaceflight that would overcome the budgetary, technological, and policy constraints of prior planning efforts. Such a new paradigm was designed to untether national efforts to go beyond low-Earth orbit, where humans had been stuck since the last Apollo lunar mission in 1972.

The following pages document internal NASA planning efforts for long-term human space exploration, as well as the negotiations between government officials at NASA and the Executive Office of the President that preceded the announcement of the Vision for Space Exploration. The first chapter after this introduction provides a sense of the ideas and concepts available to policy-makers and exploration planners at the end of the 20th century. Subsequent chapters explain the work of DPT and NEXT through the Clinton and George W. Bush administrations, including the exploration concepts the two groups developed and how they proposed dealing with difficult issues concerning the design of space missions and the long-term goals of the Agency. Later chapters explain the impact of the Columbia accident on planning at NASA and the

---

10. For background on Kennedy's famous "urgent needs" speech of 25 May 1961, see *http://history.nasa.gov/moondec.html* (accessed 1 May 2013), and for information on NASA engineer John Houbolt's early thinking about how best to land on the Moon, see James R. Hansen, *Enchanted Rendezvous: John C. Houbolt and the Genesis of the Lunar-Orbit Rendezvous Concept*, Monographs in Aerospace History no. 4 (Washington, DC: NASA, 1995), available at *http://history.nasa.gov/monograph4.pdf*. Before his speech, Kennedy famously asked Vice President Lyndon Johnson, "What can we do to beat the Russians?" For information on the policy window concept, see John W. Kingdon, *Agendas, Alternatives, and Public Policies* (New York: HarperCollins College Publishers, 1995), cited in Thor Hogan, *Mars Wars: The Rise and Fall of the Space Exploration Initiative* (Washington, DC: NASA SP-2007-4410, 2007), p. 3. Thanks to Audrey Schaffer for mentioning this useful policy window concept in a different context.

Cover slide of a DPT presentation 14 months after establishment of the team. (NASA)

White House, debates over the content of the VSE, and the early implementation of the program. We end our main story with the Shuttle's Return to Flight after the Columbia accident and some related issues during the Bush administration. The final chapter offers some summary conclusions.

Because this is a history of the policy formulation of the VSE, we do not address the shifts in space policy that occurred during administrations after President George W. Bush. We leave this subject for other historians to analyze in the future.

2
CONTEXT AND BACKGROUND OF EXPLORATION PLANNING

GIVEN THE DEARTH of potential accessible locations for humans near Earth, scenarios for human space exploration typically have included only three destinations: Earth-orbiting space stations, the Moon, and Mars.[1] Despite these limited options, plans for human exploration have been elaborate and numerous, reflecting the diversity of possibilities for space operations in terms of goals, logistics, transportation, and infrastructure. When individuals with the power to shape national policy took up the challenge of establishing a new human space exploration strategy, they had the benefit of this rich foundation of ideas to draw from, as well as four decades of U.S. and Russian experience with human spaceflight. Although they sought to leave their mark on history by creating an entirely new plan for exploration, the NASA and White House officials who crafted the VSE could not escape the past. The exploration strategy that NASA and the White House ultimately conceived took inspiration from the successes and failures of the space program and built upon ideas generated by proponents of human space exploration over the course of the 20th century.

Prelude to the Space Age

The visionaries who provided the first credible support for attempting spaceflight with rockets, Konstantin E. Tsiolkovskii in Russia and Robert H. Goddard in the United States, also proposed novel space technologies and concepts that went

---

1. Other locations and objects, including near-Earth asteroids and libration points, have received far less attention.

far beyond what could be brought to fruition in their lifetimes.[2] In his 1903 paper, "Exploration of the Universe with Rocket-Propelled Vehicles," Tsiolkovskii put forth a mathematical formula that identified the basic technical parameters for reaching and flying in space with rockets. In addition to this seminal achievement and his propositions regarding the design of rockets, Tsiolkovskii offered concepts for technologies that could support human travel in space, including space stations, spacesuits, and life-support systems and techniques. He also developed a multistage plan for space exploration and published fictional stories that included space stations, satellites, multistage rockets, and Space Shuttle–like winged gliders.[3]

Undated artist's depiction of Konstantin Tsiolkovskii. (NASA)

While Tsiolkovskii restricted himself to theory and speculation, Goddard both wrote theoretical treatises and experimented with actual rockets. He devoted a great deal of time and effort to considering means for propulsion,

---

2. Both men were heavily influenced by Jules Verne's fictional stories of space travel. Richard S. Lewis, *From Vinland to Mars: A Thousand Years of Exploration* (New York: Quadrangle, the New York Times Book Co., Inc., 1978), pp. 113–114; Frank H. Winter, *Rockets into Space* (Cambridge, MA: Harvard University Press, 1990), pp. 7–13; and Howard E. McCurdy, *Space and the American Imagination* (Washington, DC: Smithsonian Institution Press, 1997), pp. 13–18.

3. For a biography focusing on Tsiolkovskii's role in Russian/Soviet society and early public notions of spaceflight, see James Andrews, *Red Cosmos: K. E. Tsiolkovskii, Grandfather of Soviet Rocketry* (College Station, TX: Texas A&M Press, 2009). Tsiolkovskii's 1926 "Plan To Conquer Interplanetary Space" is included in a translation of some of his works, K. E. Tsiolkovskii, *Works of K.E. Tsiolkovskii on Rocket Technology—Space Investigations by Reactive Devices, Cosmic Ships and Rockets, Airplanes, Rocketplane, Fuel for Rockets, and Semi-Reactive Stratoplane* (NASA TT-F-243, November 1965, Document ID 19650027274), deposited in the NASA HRC. In this translation, the plan is covered on pp. 208–217 and includes 15 steps from a winged rocket plane to humans moving out of our solar system once the Sun starts to die. Other secondary sources translate his plan into 16 steps. See, for example, *https://web.archive.org/web/20110409065907/www.informatics.org/museum/tsiol.html* (accessed 14 June 2018).

Dr. Robert H. Goddard tows his rocket to the launching tower behind a Model A Ford truck, 15 miles northwest of Roswell, New Mexico, circa 1930–1932. (NASA 74-H-1210)

including solar energy, nuclear power, guns, and ion motors. In his personal notebooks, Goddard speculated about the possibility of sending cameras to photograph distant planets and communicating across the solar system. He also considered methods for protecting spacecraft from meteoroid debris. He is most famous for the ideas he developed in his 1919 report, *A Method of Reaching Extreme Altitudes*, and for launching the first liquid-fueled rocket in 1926.[4]

Born in Romania but a German by nationality, Hermann Oberth is considered one of the three fathers of spaceflight, along with Tsiolkovskii and Goddard. While his dissertation on rocketry and spaceflight was rejected as too speculative, he adapted it into his 1923 classic, *The Rocket into Planetary Space*, which provided mathematical support for using rockets for space travel. Oberth foresaw the use of ion propulsion and electric rockets in his 1929 book *Paths to Space Travel*, which was an expanded version of *The Rocket into Planetary Space*.[5] Oberth's books and ideas inspired space advocates Max Valier and Willy Ley,

---

4. Lewis, *From Vinland to Mars*; Winter, *Rockets into Space*, pp. 13–34; and McCurdy, *Space*, p. 16.
5. Roger D. Launius, *Frontiers of Space Exploration* (Westport, CT: Greenwood Press, 1998), pp. 87–89; Michael L. Ciancone, *The Literary Legacy of the Space Age: An Annotated Bibliography of Pre-1958 Books on Rocketry & Space Travel* (Houston: Amorea Press, 1998), pp. 47–48; and Michael Neufeld, *Von Braun: Dreamer of Space, Engineer of War* (New York: Alfred A. Knopf, 2007).

who produced books on spaceflight, served as advisors to the popular Fritz Lang film *Frau im Mond (The Woman in the Moon)*, and cofounded the German Verein für Raumschiffahrt (VfR) (German Society for Space Travel), one of the world's first significant private organizations devoted to spaceflight.[6]

On the eve of World War II, the young Briton Arthur C. Clarke published an article titled "We Can Rocket to the Moon—Now," beginning what would prove to be a prolific career writing about space.[7] Several years later, in *The Exploration of Space*, Clarke outlined a sequential program for space exploration, with an ambitious focus on humans going to multiple planets and beyond. Clarke proposed the following steps:

Hermann Oberth (forefront) with officials of the Army Ballistic Missile Agency in Huntsville, Alabama, on 27 February 1956. Seated behind Oberth are Dr. Ernst Stuhlinger and Wernher von Braun; standing behind them are Major General H. N. Toftoy and Dr. Robert Lusser. (NASA 9131100)

1. Sending robotic spacecraft around the Earth, to the Moon, and to other planets.
2. Implementing crewed suborbital rocketry.
3. Conducting piloted Earth-orbit spaceflight.
4. Sending astronauts to orbit the Moon.
5. Landing astronauts on the Moon.
6. Developing technology to refuel rockets in flight for human trips to Mars and Venus.
7. Landing humans on Mars and Venus.

---

6. Tom D. Crouch, "Willy Ley: Chronicler of the Early Space Age," in *Realizing the Dream of Flight: Biographical Essays in Honor of the Centennial of Flight, 1903–2003*, ed. Virginia P. Dawson and Mark D. Bowles (NASA SP-2005-4112, 2005), pp. 156–157.

7. McCurdy, *Space*, p. 21. Clarke's article about going to the Moon was published in *Tales of Wonder* 7 (summer 1939): 84–88. In 1945, Clarke presciently proposed the idea of geosynchronous communications satellites in "Extra Terrestrial Relays: Can Rocket Stations Give World Wide Radio Coverage?" *Wireless World* 51 (October 1945): 305–308.

After achieving these ambitious goals, humans would then travel to the outer planets and their moons and interact with intelligent life beyond our solar system.[8]

Von Braun and His Critics

After World War II, Wernher von Braun and many members of his German rocket team came to the United States under "Project Paperclip." The Army interned von Braun from 1945 to 1950 at the White Sands Proving Ground, New Mexico. He learned English during this time and wrote a novel, *The Mars Project*, in which he laid out his vision for a Mars expedition. Published in 1953, the book proposed a massive effort entailing 10 large spaceships and 70 crewmembers. The expedition would include a 400-day stay on the Martian surface and take a total of 2 years and 239 days to complete. Constructing the flotilla in Earth's orbit prior to departing for Mars would require 950 ferry rides from

This 1953 book (University of Illinois Press) was originally published in German as *Das Marsprojekt: Studie Einer Interplanetarischen Expedition* (*The Mars Project: Study of an Interplanetary Expedition*, 1952, Umschau Publishers). (Image courtesy of Michael Ciancone)

Earth's surface. While others have theorized that the large polar expeditions of the early Cold War influenced von Braun's assumptions about the resources and size of the expedition needed for a Mars journey, he claimed that his inspiration came from further back in history. As he explained in the introduction to *The Mars Project*, interplanetary exploration "must be done on a grand scale." The Mars crew would need to be highly self-reliant, as Christopher Columbus and his crew had been on their journey to the New World.[9] To properly carry out the mission for the greatest benefit to humankind, the crew would need to possess a broad range of skills and experiences.

8. Arthur C. Clarke, *The Exploration of Space* (New York: Harper and Brothers, 1951), pp. 61–62, 182, cited in McCurdy, *Space*, pp. 34–35.

9. Wernher von Braun, *The Mars Project* (Urbana, IL: University of Illinois Press, 1953), p. 2. This book was published first in German in 1952.

Before this book was published in English, von Braun wrote a series of articles for *Collier's* magazine, a periodical with a broad popular circulation at the time. In the first of eight articles, published in March 1952, von Braun outlined for a general audience the large-scale expedition approach that he explained in greater technical detail in *The Mars Project*. While assuming that humans would reach Mars without robotic precursors, von Braun's *Collier's* series added a major new feature: an Earth-orbiting space station that would be built as a way station for humans to go to the Moon. The *Collier's* series led directly to a series of television programs on space exploration produced by Walt Disney and shown on the ABC network between 1954 and 1957. During the 1950s, von Braun also collaborated with artist Chesley Bonestell to produce *The Exploration of Mars*, a richly illustrated book that included specific steps to send humans to Mars.[10]

The German rocket engineer's approach became known as the "von Braun paradigm." This paradigm entailed six elements completed in sequential order:

1. Robotic Earth-orbiting spacecraft.
2. Human spaceflight around Earth.
3. Winged reusable spacecraft.
4. A permanently inhabited Earth-orbiting space station.
5. Human lunar exploration.
6. Human exploration of Mars.[11]

A core idea of the paradigm was that humans would build up their spacefaring capabilities gradually, following this predetermined sequence.

Von Braun biographer Mike Neufeld distilled this paradigm to four components: space shuttle, space station, humans to the Moon, and humans to Mars. This paradigm has remained the de facto approach to space exploration for supporters of human spaceflight; Neufeld called it an "entrenched mindset at NASA" for the last half century. Neufeld also pointed out differences between von Braun's personal views and the more intellectually tidy paradigm that bears his name. Although he initially assumed that a station would be built before humans would explore either the Moon or Mars, for example, von Braun did

---

10. Wernher von Braun, "Crossing the Last Frontier," *Collier's* (22 March 1952); David S. F. Portree, *Humans to Mars: Fifty Years of Mission Planning, 1950–2000*, Monographs in Aerospace History no. 21 (Washington, DC: NASA SP-2001-4521, 2001), p. 2; and McCurdy, *Space*, p. 38.

11. See Dwayne Day, "The Von Braun Paradigm," *Space Times* (November–December 1994): 12–15; and Roger D. Launius, "First Steps into Space: Projects Mercury and Gemini," in *Exploring the Unknown: Selected Documents in the History of the U.S. Civil Space Program, Volume VII: Human Spaceflight: Projects Mercury, Gemini, and Apollo*, ed. John M. Logsdon with Roger D. Launius (Washington, DC: NASA SP-2008-4407), pp. 2–4.

not believe that a space station was a necessary step for sending astronauts to Mars.[12]

Neufeld contended, furthermore, that von Braun was at least as interested in the Moon as in Mars. In a 1970 interview, von Braun expressed the belief that he had been inaccurately maligned as a "Mars or bust" advocate. Despite the Moon's allure for von Braun, he understood the Western cultural resonance of Mars. He felt that conducting a human mission to the Moon would be too easy, so he often emphasized Mars publicly and is remembered more as a Mars advocate than a lunar advocate. Despite his utopian desire to alter the course of humanity through space exploration, von Braun was a pragmatist when it

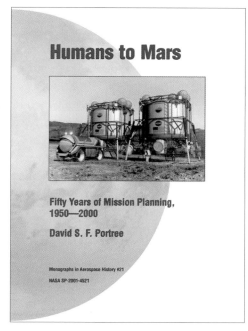

Humans to Mars

Fifty Years of Mission Planning, 1950—2000

David S. F. Portree

Monographs in Aerospace History #21
NASA SP-2001-4521

Published in 2001, David Portree's *Humans to Mars* monograph analyzes 50 of the most important concept design studies conducted between 1950 and 2000. (NASA)

came to national politics. He understood that after the Apollo triumphs, neither Congress nor the White House would be inclined to allocate the tremendous financial resources required to send humans to Mars.[13]

Von Braun's ideas were highly influential. In *The Exploration of Mars*, author and illustrator Ron Miller noted that "von Braun had developed a detailed, consistent, incremental plan for the exploration of space. In much the same way that Jules Verne had shown a century earlier that spaceflight was merely a matter of applied mathematics and engineering," von Braun convinced policy-makers

12. See Michael J. Neufeld, "The 'Von Braun Paradigm' and NASA's Long-Term Planning for Human Spaceflight," chap. 13 in *NASA's First 50 Years: Historical Perspectives*, ed. Steven J. Dick (Washington, DC: NASA SP-2010-4704, 2010), pp. 325–347.

13. Neufeld, "Von Braun Paradigm." The von Braun quotation is from von Braun's interview with John Logsdon, 25 August 1970, pp. 4 and 3, file 2629, NASA HRC, cited in Neufeld, "Von Braun Paradigm," p. 340. See also Neufeld, *Von Braun: Dreamer of Space, Engineer of War*, pp. 400, 447.

and the public that human spaceflight was technically feasible, worthwhile, and exciting.[14]

Despite the great influence of the von Braun paradigm on the space program, NASA has not followed the more rigid version of the paradigm over the years. For example, President Kennedy's decision to send astronauts directly to the Moon to overtake the Soviet Union in the space race ensured that an Earth-orbiting space station would not precede a lunar landing. The deployment of the reusable Space Shuttle prior to the development of a continuously inhabited, large-scale U.S. space station also diverged from the paradigm. Von Braun also failed to anticipate in the early 1950s the possibility that revolutionary advances in information and communications technologies could reduce the personnel requirements for a human mission to Mars. Nor did he envision the extent to which advanced avionics, information technology, and telecommunications would enable automatic control of spacecraft and the blossoming of a vibrant robotic planetary exploration program.[15] None of these points minimizes von Braun's influence on the human spaceflight agenda. He did not believe that the steps he laid out for human space missions in his public writings needed to be followed in strict order. His overriding goal was advancing the cause of human spaceflight, both politically and technologically.

The von Braun paradigm was far from the only logical approach to space exploration. Although his ideas are critical for understanding the direction of the human space program, von Braun focused narrowly on sending humans to new places and ignored less dramatic opportunities for deepening scientific understanding of the physical and biological universe. He viewed a space station, for example, as dedicated infrastructure that would be used as a stepping-off point for planetary exploration and in-space assembly of rockets. His Soviet contemporaries, on the other hand, viewed space stations as less permanent and of more limited use. Von Braun also did not envision that a space station might be used as a scientific research platform, as the Soviets did with their Mir station and participating nations have done with the International Space Station.[16]

Proponents of alternative approaches to exploration occupied key positions in the major space programs of the 1950s. Milton Rosen, the leader of Viking

---

14. Ron Miller, "Spaceflight and Popular Culture," in *Societal Impact of Spaceflight*, ed. Steven J. Dick and Roger D. Launius (Washington, DC: NASA SP-2007-4801, 2007), p. 510.

15. Von Braun, *The Mars Project*, p. 75, cited in Portree, *Humans to Mars*, p. 1. Portree, p. 2, cites Louise Crossley, *Explore Antarctica* (Cambridge, England: Cambridge University Press, 1995), p. 40, in noting that Operation High Jump to Antarctica in 1946–1947 entailed 13 ships, 23 airplanes, and 4,000 personnel. See Portree, *Humans to Mars*, p. 2, for these points about communications advances and manual vs. automatic control of spacecraft.

16. Day, "The Von Braun Paradigm," pp. 14–15.

rocket development and the technical director of the Vanguard project that followed, favored robotic over human space exploration, partly due to the meager state of knowledge about space during his time. He believed that sending people into space would be extremely difficult, considering all of the systems integration, systems reliability, human factors, and radiation challenges.[17] James Van Allen, the scientist behind the U.S.'s first successful spacecraft, Explorer 1, for many years vocally opposed human spaceflight in favor of robotics.[18] Whereas von Braun saw space as an empty place that was destined to be explored and settled by humans, Rosen and Van Allen favored robotic exploration because they believed that the physical barriers to maintaining human life in space were too great. The possibilities for maintaining a long-term human presence in space were small, Rosen and Van Allen believed, given the harsh conditions of space and the extensive artificial apparatuses needed merely to allow for survival.[19]

Concerned about the risks and budgetary impact of human spaceflight, President Eisenhower was more inclined toward Rosen and Van Allen's points of view. Eisenhower preferred a relatively small civil space program based upon scientific study with robotic craft. He supported Project Mercury because he was convinced of the benefits of learning more about the impact of microgravity on the human body and the feasibility of human operations in space. Eisenhower resisted the fledgling Project Apollo because he felt that it lacked a sufficiently strong scientific basis and would be a mere publicity stunt.[20]

From Sputnik to Challenger

In the wake of the Soviet Union's Sputnik spectacular in October 1957, the National Advisory Committee for Aeronautics (NACA) established a group to consider how to go into space and tapped von Braun to chair it. In July 1958,

---

17. William J. Laurence, "Two Rocket Experts Argue 'Moon Plan,'" *New York Times* (14 October 1952), cited in Launius and McCurdy, *Robots in Space*, pp. 65–66.

18. See, for example, James Van Allen, "Space Station and Manned Flights Raise NASA Program Balance Issues," *Aviation Week and Space Technology* (25 January 1988): 153 (cited in Launius and McCurdy, *Robots in Space*, p. 67); and James A. Van Allen, "Is Human Spaceflight Obsolete?" in *Issues in Science and Technology* 20, no. 4 (summer 2004), available at *http://www.issues.org/20-4/p_van_allen/* (accessed 16 April 2018). Roger Launius has grouped together the ideas of Rosen, Van Allen, and President Dwight Eisenhower and labeled them the Rosen–Eisenhower–Van Allen paradigm. Simply understanding that there were multiple alternative concepts for spaceflight at odds with the ideas von Braun promoted is sufficient for the purposes of this study. On the notion that the alternatives formed a coherent paradigm, see Roger Launius, "Sphere of Influence: The Sputnik Crisis and the Master Narrative," *Quest: The History of Spaceflight Quarterly*, 14, no. 4; and Launius and McCurdy, *Robots in Space*.

19. Launius and McCurdy, *Robots in Space*, p. 66.

20. Ibid., pp. 66–67.

the working group came up with a detailed, sequential plan with established milestones for the United States to continue its robotic spaceflight program and send humans into space. The plan called for the United States to launch progressively larger payloads into Earth orbit and to the Moon; learn to send humans safely into Earth orbit to live in space stations; and then go to the Moon and on to other planets. In retrospect, the plan was extraordinarily ambitious: a space station with a crew of four by 1962, a 20-person space station by 1964, humans landing on the Moon in 1966, establishment of a permanent lunar base by 1974, and a human expedition to a planet (presumably Mars) by 1977.[21]

Shortly after NASA's establishment in October 1958, the new "space agency" crafted a similar sequential long-range plan, but with a less aggressive schedule than the one von Braun's group had proposed. The 1959 plan called for suborbital flights for piloted spacecraft, orbital flights, a trip around the Moon, a lunar landing, robotic flights to Mars and Venus, and then astronauts flying to those two planets. The long-range planners demurred on a large space station but did include a smaller one.[22]

Even while Apollo development efforts were accelerating in the early 1960s, NASA's Future Projects Office sponsored a series of Early Manned Planetary-Interplanetary Roundtrip Expeditions (EMPIRE) studies.[23] NASA's Office of Manned Space Flight also considered future piloted planetary missions, including flybys of Mars. Shortly after taking over as the head of this organization in the fall of 1963, George Mueller set up an Advanced Manned Missions Office to oversee such planning work. Less than two years later, he received permission from Robert Seamans, the NASA Deputy Administrator, to put together a team with NASA Headquarters and Center employees to plan planetary missions. The team, later called the Planetary Joint Action Group (JAG), first focused on sending astronauts to the surface of Mars via nuclear rockets. In

---

21. NACA Special Committee on Space Technology, Working Group on Vehicular Program, "A National Integrated Missile and Space Vehicle Development Program," 18 July 1958, p. 6, cited in Edward C. Ezell and Linda N. Ezell, *On Mars: Exploration of the Red Planet, 1958–1978* (Washington, DC: NASA SP-4212, 1984), pp. 9, 10, and especially 11. This NACA Working Group document is available in file 15853 in the NASA HRC.

22. McCurdy, *Space*, p. 49, and NASA's Office of Program Planning and Evaluation, "The Long Range Plan of the National Aeronautics and Space Administration," 16 December 1959, file 17806, NASA HRC, cited in McCurdy, *Space*, p. 49. This long-range plan states on p. 28 that "Manned exploration of the moon and the nearer planets must remain as the major goals for the ensuing decade."

23. For more on the EMPIRE studies, see chapter 3 of Portree's *Humans to Mars*, pp. 11–22. Anne M. Platoff's master's dissertation, "Eyes on the Red Planet: Human Mars Mission Planning, 1952–1970" (University of Houston–Clear Lake, May 1999) also covers this time period in considerable detail. In particular, her dissertation includes separate chronological appendices on both human and robotic exploration of Mars.

1966, however, the JAG looked at a Mars flyby mission instead and Mueller testified on Capitol Hill about the possible mission.[24]

After the Apollo 204 (later renamed Apollo 1) fire in January 1967 and the riots and social upheaval of 1968, humans-to-Mars planning continued, although the Agency focused largely on the Apollo program. In late 1969, Mueller convened the Planetary Mission Requirements Group (PMRG), the successor to the Planetary JAG. The PMRG, however, did not receive high-level support amidst administration efforts to cut NASA's budget. Formal planning for a Mars expedition came to a halt with the hurried completion of the PMRG effort.[25]

Shortly after the Apollo 11 mission, the White House convened a Space Task Group to devise what the group called in its final report "a vision for the future" of the nation's space program.[26] The Space Task Group initially developed three options for building upon the achievements of the Apollo program. The options varied primarily in cost, the timing of development and operations, and the final decision point for determining whether to continue to Mars. The major elements of all three options included a Moon base, a lunar-orbiting station, a massive 50-person Earth-orbiting space station, and intentions to send humans to Mars. By the time President Richard Nixon was ready to make a decision on the program, the only choice he had to make was whether to support the development and future deployment of an Earth-orbiting space station and space shuttle, nothing at all, or one or the other. Serious steps toward a crewed Mars mission had been excised from the list of options. Significantly curtailing the ambitions of space enthusiasts, President Nixon decided to authorize the development of the Space Shuttle in January 1972.[27] During the remainder of the 1970s, interest in Mars at NASA tended to focus on the two Viking robotic spacecraft, intended to investigate the possibility of extant life on Mars, while human spaceflight enthusiasts focused their ambitions on the development of the Shuttle, which first flew in 1981.

While planning for human exploration of Mars died out in the early 1970s, support for human planetary missions occasionally arose from unexpected sources. A report of a White House Science Advisor's panel completed in March 1970, for example, called for continued robotic and human exploration

---

24. Portree, *Humans to Mars*, pp. 24–32 and especially 24–26.

25. Ibid., pp. 48–49.

26. The Space Task Group's September 1969 report is available at *http://history.nasa.gov/taskgrp.html*.

27. See T. A. Heppenheimer, *The Space Shuttle Decision: NASA's Search for a Reusable Space Vehicle* (Washington, DC: NASA SP-4221, 1999) for Nixon's Shuttle decision and *http://history.nasa.gov/stsnixon.htm* for the text of Nixon's announcement.

of the Moon in a carefully planned sequence even though the report's authors believed that robotic spacecraft were more cost-effective than astronauts for obtaining scientific information from space. Regarding Mars, the report noted that much useful information could be gleaned from robotic spacecraft alone and that more research was needed on long-duration human spaceflights. In projecting that "manned planetary exploration could become a reality before the end of the twentieth century," the report appeared to express acceptance of a future in which human missions dominated the nation's civil space program.[28]

Popular interest in human exploration beyond LEO did not abate with President Nixon's Space Shuttle decision. Advocacy groups sprang up, in part out of fear that the ambitions of the U.S. civilian space program would remain modest long into the future. Many of these private groups had a small membership base and focused narrowly on promoting particular approaches to exploration or travel to specific destinations in space.[29] These groups, nonetheless, carried the torch for more ambitious human space missions than NASA could accomplish with the Space Shuttle. While NASA focused on the extraordinary challenges of building and flying the first fully functional Space Shuttle, private organizations focused on stimulating public interest in space. Two of the major groups formed in the 1970s, Wernher von Braun's National Space Institute and the L5 Society (which was inspired by Princeton University professor Gerard O'Neill's ideas about space colonization), attracted national attention, partly due to the celebrities associated with the groups. The National Space Institute's board of directors, for example, included Buzz Aldrin (astronaut), Jacques Cousteau (explorer), John Glenn (astronaut and politician), Bob Hope (entertainer), and Gene Roddenberry (TV producer).[30]

The nonprofit L5 Society proved more productive in terms of inspiring futurists and scientists to develop concepts ultimately intended to bring the

28. Executive Office of the President, Office of Science and Technology, "The Next Decade in Space: A Report of the Space Science and Technology Panel of the President's Science Advisory Committee," March 1970, pp. ii, 2. A copy of this report is available in file 12450 in the NASA HRC.

29. See, for example, Amy Paige Snyder, "What Place in Space? Impacts and Future Prospects of the U.S. Pro-Space Movement," *Ad Astra* (March/April 2001): 44–47; and Michael A. G. Michaud, "The Beginnings of the New Space Movement," *L5 News* (June 1985), available at *http://www.nss.org/settlement/L5news/1985-beginnings.htm*, accessed 16 April 2018.

30. Linda Billings, "Overview: Ideology, Advocacy, and Spaceflight—Evolution of a Cultural Narrative," in *Societal Impact of Spaceflight*, ed. Steven J. Dick and Roger D. Launius (Washington, DC: NASA SP-2007-4801, 2007), p. 490; Richard Godwin, "The History of the National Space Society," *http://www.space.com/adastra/adastra_nss_history_051116.html* (accessed 18 July 2008).

This space habitat design represents concepts developed by Princeton University professor and Mars advocate Gerard O'Neill in the 1970s. This particular one, featuring a wheel over a mile in diameter, resulted from a NASA Ames Research Center (ARC)–Stanford University summer study. (Courtesy of the Space Studies Institute, *http://ssi.org/space-art/ssi-sample-slides/*)

dream of human space settlements to reality.[31] Space enthusiasts Keith and Carolyn Henson founded the L5 Society[32] in August 1975, less than a year following the publication of O'Neill's September 1974 *Physics Today* article, "The Colonization of Space." Carolyn Henson arranged for political support from Morris Udall, an Arizona Congressman running for president. Robotics pioneer Hans Moravec, futurist Eric Drexler, artificial intelligence expert Marvin Minsky, and countercultural figure Timothy Leary joined early on. Science fiction pioneers Isaac Asimov and Robert Heinlein and noted physicist Freeman Dyson became L5 directors.[33]

In 1977, O'Neill began efforts to establish a separate nonprofit group to maintain support for research into developing space colonies. The next year, the Space Studies Institute began operations from a tiny office on Princeton's campus. This institute undertook cooperative technical research projects with

31. "Space Studies Institute—History," *http://ssi.org/about/history/* (accessed 26 October 2011).

32. L5 is one of five Lagrangian points where the gravitational pulls of Earth and the Moon and centrifugal force are in equilibrium. O'Neill and the L5 Society were enthusiastic about creating whole terrestrial Earth-like colonies in space there.

33. Ed Regis, *Great Mambo Chicken and the Transhuman Condition: Science Slightly Over the Edge* (Reading, MA: Addison-Wesley Publishing, 1990), p. 62.

academic and industrial partners. An enthusiastic proponent of human space exploration who helped spark popular interest in space colonization, O'Neill has been recognized as one of the more important, if less well known, promoters of space exploration in the late 20th century. His son, Roger, took over leadership of the Space Studies Institute after Gerard O'Neill's death in 1992.[34] The organization inspired by O'Neill, the L5 Society, merged with the National Space Institute in 1987 to form the more widely known National Space Society (NSS), which is an "independent, educational, grassroots, non-profit organization dedicated to the creation of a spacefaring civilization."[35]

Other equally important space advocacy groups and research-oriented organizations emerged in the 1980s and 1990s to promote various aspects of space exploration, including the Planetary Society in 1980, to promote robotic exploration; the Students for the Exploration and Development of Space in 1980, to promote space exploration generally; and the Space Access Society in 1992, to encourage cheap and routine access to space. While such groups have grown larger in size and more numerous over the years, only the National Space Institute and L5 Society filled the gap left when both the Apollo program ended and planning for human planetary missions came to a halt in the early 1970s.

The first serious attempt to rekindle planning for long-term human planetary exploration fell victim to bad timing. The National Commission on Space, known as the Paine Commission for its chairman, former NASA Administrator Thomas Paine, began a yearlong study in March 1985 to produce a plan for space exploration into the 21st century. Congress created this independent Commission and gave the President authority to appoint its members, which included notable figures in the space community such as Neil Armstrong (astronaut), Chuck Yeager (pilot), and Gerard O'Neill, as well as Ambassador Jeane Kirkpatrick and retired Air Force General Bernard Schriever. The final report, titled *Pioneering the Space Frontier*, offered a bold plan for human exploration of the Moon and Mars and for human settlement of space. Presented to President Ronald Reagan in July 1986, the report had been commercially published two months earlier. It had almost no influence on NASA because the Agency was then focused almost exclusively on one of the biggest tragedies in its history. The Commission made no adjustments to the report to account for the possible political or technical impact of that tragedy.[36]

---

34. "Space Studies Institute—History."

35. "About the National Space Society (NSS)," *http://www.nss.org/about/* (accessed 6 February 2008).

36. See *Pioneering the Space Frontier: An Exciting Vision of Our Next Fifty Years in Space, The Report of the National Commission on Space* (New York: Bantam Books, May 1986) is also available from *http://history.nasa.gov/painerep/cover.htm*. The point about poor timing is from Neufeld's

Challenger

The January 1986 Challenger Space Shuttle accident focused attention on the direction of NASA's human spaceflight program. The presidentially chartered Rogers Commission investigated the proximate cause of the accident and recommended specific technical solutions to prevent future breaks in the seals of the by-then-infamous O-rings.[37] At the same time, debate emerged within the Agency regarding the extent to which the Shuttle could be flown routinely.

After the Rogers Commission completed its work in the summer of 1986, NASA Administrator James Fletcher tapped astronaut Sally Ride to lead an internal NASA study group to develop yet another plan for the Agency's future. Published in August 1987, the "Ride Report" (officially titled *NASA Leadership and America's Future in Space*) recommended carrying forth a sequential exploration program via a "strategy of evolution and natural progression." The report had four specific recommendations: creating an ambitious program of robotic solar system exploration with a focus on Mars rovers and sample return missions, establishing a large Mission to Planet Earth (MTPE) (Earth science) program, sending humans to Mars, and creating a human outpost on the Moon. In its conclusion, the report suggested increasing "our capabilities in transportation and technology, not as goals in themselves, but as the necessary means to achieve our goals in science and exploration."[38]

The Ride Report was notable for reinvigorating high-level interest at NASA in science, particularly Earth science. The report also provided the impetus for Fletcher to establish NASA's Office of Exploration in June 1987 and name Ride as the Acting Assistant Administrator for Exploration. Although the Office of Exploration supported a wealth of research on Mars exploration, the Ride Report's recommendations regarding a human lunar outpost and sending astronauts to Mars did not directly influence the course of the human spaceflight

---

"Von Braun Paradigm" chapter, p. 341. On the next page, Neufeld also criticizes this report as contributing to the "flavor of impractical utopianism."

37. *Report of the Presidential Commission on the Space Shuttle* Challenger *Accident* (Washington, DC: 6 June 1986). The full-text of this multi-volume report is available at *http://history.nasa.gov/rogersrep/51lcover.htm*.

38. The Ride Report is available at *http://history.nasa.gov/riderep/cover.htm* (accessed 16 April 2018). See also its introduction section (*http://history.nasa.gov/riderep/intro.htm*) and its "Evaluation of Initiatives" section (*http://history.nasa.gov/riderep/evalaut.htm*). The two quotations are from the conclusions section (*http://history.nasa.gov/riderep/conclus.htm*). For more on the Office of Exploration, see the organizational charts available at *http://history.nasa.gov/orgcharts/orgcharts.html#1980*; files nos. 18148–18150 in the NASA HRC; and "NASA Establishes Office of Exploration," NASA News Release 87-87, 1 June 1987, cited in Howard E. McCurdy, "The Decision To Send Humans Back to the Moon and on to Mars," NASA HHR-56, March 1992, p. 10, available in the NASA HRC.

program. Among other possible reasons, there was little political imperative then to commit to long-range plans in space that diverged from the goals of the Shuttle program.[39]

Space Exploration Initiative

On 20 July 1989, President George H. W. Bush stood on the steps of the National Air and Space Museum on the 20th anniversary of the Apollo 11 lunar landing and declared his intention to have NASA build what was then known as Space Station Freedom, send humans back to the Moon, and then ultimately send astronauts to Mars.[40] This plan, known as the Space Exploration Initiative (SEI), was designed to be a long-term endeavor with a 30-year time horizon, as opposed to an Apollo-style crash program. In response to a question at a luncheon that day, Richard Darman, the director of the Office of Management and Budget and a self-confessed space enthusiast, estimated $400 billion as a rough ballpark for SEI.[41] President Bush asked Vice President Dan Quayle, the head of the National Space Council, to identify the funding, personnel, and tools that would be required to achieve SEI's goals. Quayle asked NASA Administrator Richard Truly to help define the program with greater precision. Truly, in turn, asked Johnson Space Center (JSC) Director Aaron Cohen to lead a study team. Cohen's 90-Day Study team produced a report estimating that it would cost approximately $500 billion spread over 20–30 years.[42]

The study team's results faced several lines of criticism. First was the sticker shock of the half-trillion-dollar budget estimate. In addition, Mark Albrecht, the executive director of the National Space Council, was angry that despite his proddings, Cohen and his team had failed to come up with different technical

---

39.　Portree, *Humans to Mars*, pp. 69–73.

40.　On the history of space stations, including Space Station Freedom, see Roger D. Launius, *Space Stations: Base Camps to the Stars* (Washington, DC: Smithsonian Books, 2003).

41.　Richard Darman, remarks at a National Press Club luncheon, 20 July 1989, p. 5, transcription from the Federal Information Systems Corporation/Federal News Service via LexisNexis and copy in file 429 in the NASA HRC. Former National Space Council Director Mark Albrecht recollects that in the policy run-up to the President's formal speech, he, Darman, and others estimated the cost of SEI as ranging anywhere from $100 to $400 billion. See Albrecht, *Falling Back to Earth*, pp. 33, 39, 44, 49.

42.　For more information about SEI, see Thor Hogan, *Mars Wars: The Rise and Fall of the Space Exploration Initiative* (Washington, DC: NASA SP-2007-4410, 2007). Links to this and other SEI information such as Cohen's "90-Day Study" and brief SEI summaries by Hogan and former NASA Chief Historian Steven J. Dick are available at *http://history.nasa.gov/sei.htm* (accessed 16 April 2018). For an interesting analysis of these SEI budget estimates and their ostensible origins, see Dwayne A. Day, "Aiming for Mars, Ground on Earth: Part Two," *http://www.thespacereview.com/article/106/9*, 23 February 2004 (accessed 28 August 2014).

President George H. W. Bush speaks on the steps of the National Air and Space Museum on 20 July 1989, the 20th anniversary of the Apollo 11 Moon landing. Bush announced his new Space Exploration Initiative to complete what was then called Space Station Freedom, return astronauts to the Moon, and bring Americans to Mars for the first time. (NASA 89-H-380)

approaches or architectures to safely send astronauts back to the Moon and on to Mars. Albrecht wanted alternative policy options, but Cohen resisted any plan that sacrificed safety when the lives of astronauts were at stake. In frustration, Albrecht brought in a team from Lawrence Livermore National Laboratory to brief the National Space Council on an alternative, much less expensive plan.[43] Whether the Livermore team, led by astrophysicist Lowell Wood, really had the specialized expertise to devise such a plan is unclear.

One analyst criticized the 90-Day Study as "essentially...a shopping list of every program that various NASA constituencies wanted, regardless of whether they were necessary" or too expensive. The study also did not distinguish between necessary and desirable technologies.[44] Three other factors arguably drove up the cost estimates: increasing the time frame for SEI, increasing the financial reserves, and increasing the complexity. Adapting an automobile analogy slightly, Dwayne Day contended that the 90-Day Study "offered five different ways of doing essentially the same massive and expensive mission—like

43.   Hogan, *Mars Wars*, pp. 95–96. This episode was caused by and further eroded poor relations between NASA and the National Space Council at the time.
44.   Dwayne A. Day, "Aiming for Mars, Ground on Earth: Part Two."

selecting five different ways of paying for a Rolls-Royce, rather than looking at cheaper cars."[45]

Another, more subtle, criticism of the 90-Day Study was of its methodology, in particular its use of parametric cost estimating. Briefly, this technique uses relevant historical data in conjunction with identifying known parameters for a new situation. The rub was that while NASA had sent astronauts to the Moon during the Apollo program, no organization had ever seriously attempted, let alone succeeded, at sending people to Mars.[46]

Cohen's team used three parametric cost models—from NASA's Jet Propulsion Laboratory (JPL), JSC, and Marshall Space Flight Center (MSFC)—that relied on "historical NASA programs." The MSFC model was used for estimating space transportation vehicle costs, the JSC model for surface systems, and the JPL model for associated robotic space science efforts. These three models were then wrapped up into two overall "reference approaches," which were essentially high and low estimates based on mission timelines. Using the first reference approach, the team came up with an overall figure of $541 billion (in 1991 dollars, spread over more than 30 years). The second reference approach yielded an estimate of $471 billion.[47] Hence, most observers simply averaged these to approximately a half trillion dollars.

In his memoir, Albrecht wrote of studies using existing Soviet launch vehicles that could bring SEI's cost down to $100 billion and noted that the National Space Council "settled on a total program cost estimate of $200 billion." He also recounted a story of Darman informing John Sununu, the White House Chief of Staff, that SEI would cost $200–$300 billion. Unfortunately, he did not provide any documentation or citations for these figures.[48] Hogan noted that a major weakness of the 90-Day Study team report was the lack of "alternatives that were based on significantly different mission profiles or technical systems," which was compounded by a sense that nobody else seemed to have any other, solid cost estimates for such an ambitious human spaceflight program.[49]

---

45. Ibid.

46. Special thanks to Zach Pirtle for this key insight. One basic way to estimate cost for future spacecraft missions is based on two key factors: mass and power of the systems on board. This works reasonably well for robotic scientific missions. Thanks to Craig Tupper for his help with this aspect. But such a methodology would have been less helpful when the human spaceflight systems remained relatively undefined.

47. NASA, "Report of the 90-Day Study on Human Exploration of the Moon and Mars: Cost Summary," November 1989, available at *https://history.nasa.gov/90_day_cost_summary.pdf* (accessed 17 April 2018). See p. 2 for the brief quote.

48. Albrecht, *Falling Back to Earth*, pp. 33, 44. Albrecht also notes (incorrectly) on p. 49 that Cohen's team stuck with a $400 billion figure.

49. Hogan, *Mars Wars*, p. 91.

While the National Academy of Sciences "largely concurred with the NASA study, White House and congressional reaction to the NASA plan was hostile, primarily due to the cost estimate."[50] Because of the public and congressional chagrin over the half-trillion-dollar estimate, the White House asked for several other studies to validate the 90-Day Study's approach.

In 1990, another blue-ribbon panel, the Advisory Committee on the Future of the U.S. Space Program, was created to address the diffuse civil space program's goals in the wake of the Challenger accident, the spherical aberration problem of the Hubble Space Telescope (HST), other technical problems with the Shuttle, and the National Space Council's dissatisfaction with NASA's response to the SEI plan. This commission became more widely known as the Augustine Commission after its chairman, Norman Augustine, who had previously served as the chairman of the Lockheed Martin Corporation and as Under Secretary of the Army. This panel of outside experts reported both to Vice President Dan Quayle, in his role as Chairman of the National Space Council, and to NASA Administrator Richard Truly. The report, issued in December 1990, rejected the idea of drastically rescoping NASA's mission to focus exclusively on a new Mars exploration initiative or turning over Space Shuttle operations to another organization. The Committee argued for an invigorated NASA role in administering the U.S. civil space program.[51]

The Augustine Commission emphasized that the robotic "space science program warrants highest priority for funding" above even human spaceflight because "science gives vision, imagination, and direction to the space program." The Commission also called for a robust MTPE program, investment in new launch vehicle technology, and human spaceflight goals of a space station, lunar outpost, and robotic spacecraft that would precede astronauts traveling to Mars. The Commission also "share[d] the view of the President that the long-term magnet for the manned space program is the planet Mars—the human

---

50. Steve Dick, "Summary of Space Exploration Initiative," *http://history.nasa.gov/seisummary.htm* (accessed 17 April 2018).

51. Report of the Advisory Committee on the Future of the U.S. Space Program, December 1990, available at *http://history.nasa.gov/augustine/racfup1.htm* (accessed 17 April 2018; same for other URLs in this note). The report's Executive Summary is at *http://history.nasa.gov/augustine/racfup2.htm*, and its main recommendations are summarized at *http://history.nasa.gov/augustine/racfup6.htm*. The terms of reference (this group's charter) are at *http://history.nasa.gov/augustine/racfup7.htm#Appendix%20II*. See also the headnote of document IV-20 in John M. Logsdon et al., eds., *Exploring the Unknown: Selected Documents in the History of the U.S. Civil Space Program, Volume I: Organizing for Exploration* (Washington, DC: NASA SP-44407, 1995), pp. 741–742.

exploration of Mars, to be specific."[52] This belief, while perhaps seemingly unremarkable, represented a formal acknowledgment that the White House and space policy leaders shared the goals of space advocates who longed for signs of a national commitment to human space exploration beyond the Moon.

In terms of funding, the Augustine Commission "believe[d] that a program with the ultimate, long term objective of human exploration of Mars should be tailored to respond to the availability of funding, rather than adhering to a rigid schedule." Thus, the Augustine Report was also known for advocating a "go as you pay" approach, specifically for the MTPE program. When Norm Augustine was asked to define this go-as-you-pay approach, he replied that "when there are problems in the program, as there will always be, the schedule should be slipped rather than taking money from other smaller programs such as the research program."[53] In other words, given the traditional three elements of project management (i.e., performance, time, and money), the timetables should be the most flexible. Yet the committee report also assumed a very significant buildup in NASA funding over the next decade: 10 percent per year in real dollars.[54]

Five months later, in May 1991, another high-profile commission led by former astronaut Thomas Stafford submitted a report to Vice President Dan Quayle, the head of the National Space Council. The "Synthesis Group Report," also known as the "Stafford Report," was formally called *America at the Threshold: America's Space Exploration Initiative*. Although SEI had already experienced great political difficulties, the report still outlined an ambitious set of plans to send humans to Mars. Rather than limit the study to an internal group of engineers and program managers, as had been done with SEI's 90-Day Study, the Synthesis Group deliberately included non-NASA people to tap the creativity and fresh ideas that lay beyond the reach of the bureaucracy.[55] The group presented four architectures, the first of which, called "Mars Exploration," focused

---

52. See *http://history.nasa.gov/augustine/racfup2.htm* (accessed 17 April 2018) for these quotations and other information.

53. See *http://history.nasa.gov/augustine/racfup2.htm*, *http://history.nasa.gov/augustine/racfup5.htm*, and *http://history.nasa.gov/augustine/racfup6.htm* (all accessed 17 April 2018). The quotation from Norm Augustine is actually from the "Stafford Report" (see next paragraph), *http://history.nasa.gov/staffordrep/main_toc.pdf* (accessed 10 July 2018), p. 17. Astronaut Don Pettit defined "go as you pay" in another, simple way: "show results for what you paid for." See Don Pettit's 25 June 1999 input to DPT Phase 1, Space Architect Doc Archive files, deposited in the NASA HRC.

54. See *http://history.nasa.gov/augustine/racfup2.htm* (accessed 17 April 2018).

55. Thanks to Lisa Guerra for pointing this out. She performed some staff work for the Synthesis Group, as did Don Pettit. Pettit later became an astronaut and, like Guerra, was a member of DPT.

on sending humans to the Red Planet while allowing for some necessary preparatory activities on the Moon that would have significant ancillary scientific content. The second architecture, "Science Emphasis for the Moon and Mars," made the case for the integration of robotic and human spaceflight, presaging both DPT and the VSE. The third, "The Moon To Stay and Mars Exploration," focused on human exploration of the Moon and smaller-scale crews going to Mars. The fourth architecture, "Space Resource Utilization," emphasized the use of the Moon's resources for lunar surface operations and, potentially, to send crews on to Mars. While the four architectures differed in detail, the Synthesis Group Report called for astronauts to land on Mars between 2014 and 2019, with the initial stay on Mars to last 30–100 days and then increase to approximately 600 days for subsequent missions. Each mission to Mars would entail six crew members all descending to the Martian surface.[56]

Despite the substantial effort that went into the Synthesis Group Report, it could not alter the fate of the Space Exploration Initiative. Without any major political support for SEI outside of NASA, the initiative dissolved gradually from public view. Administrator Truly was dismissed by the Bush administration in the spring of 1992, and the NASA Headquarters Exploration Office was shuttered in late 1992.[57] The plans and ideas that came out of this two-year flurry of activity, at least for the time, took their place on the bookshelves of NASA employees and leaders comfortably alongside volumes of past studies that faced a similar fate.

Contrary to the opinions of those who viewed SEI as doomed to fail (principally because of its enormous price tag), political scientist Thor Hogan contended that SEI's failure was due to "a deeply flawed [NASA and White House] decision-making process that failed to develop (or even consider) policy options that may have been politically acceptable given the existing political environment."[58] Hogan argued that the 1970s and 1980s were times of decline for and public apathy toward NASA that left the Agency with no major supporters in Congress or the White House. A flagging economy and an increasing federal deficit when Bush took office in 1989 meant more bad news for NASA. Despite all of these challenges facing any major new space initiative (or perhaps in part because of some of these issues within NASA), "SEI reached the national agenda after an incredibly short two-month alternative generation process that

---

56. The report is available at *http://history.nasa.gov/staffordrep/main_toc.pdf* (accessed 17 April 2018), and there is a good summary of the alternative approaches that were proposed in the "Architectures" chapter, found at *http://history.nasa.gov/staffordrep/arch_pre.pdf* (accessed 17 April 2018), especially on pp. 16–17.

57. See, for example, Hogan, *Mars Wars*, passim and pp. 164–165.

58. Hogan, *Mars Wars*, p. 2.

was conducted in secret by senior leaders at NASA Headquarters and Johnson Space Center." Not only was the policy generation period remarkably short, but the Bush administration put forth the initiative only a few months after entering office.[59]

The Goldin Years

While it may have faded from public view, planning for human space exploration beyond LEO did not die with SEI. Two studies conducted in the 1990s, the Mars Reference Mission (MRM) and Human Lunar Return (HLR), continued to deepen understanding of the requirements of piloted missions and laid out detailed plans for accomplishing them. At the same time, a series of successful missions with relatively inexpensive robotic spacecraft that were designed and deployed under compressed schedules encouraged the Agency's new Administrator, Dan Goldin, to consider a more ambitious agenda. The success of the first Mars Rover (Sojourner), the experience of the Hubble Space Telescope servicing missions, and possible indications of life in the Martian past also generated greater public interest in NASA activities and transformed key leaders in NASA's science programs into human exploration advocates.[60]

As the remnants of SEI were fading away, NASA organized a working group to investigate options for space exploration and once again begin preliminary planning for sending humans to Mars. Led by JSC personnel, the team came to include representatives from many of the NASA Centers.[61] The Associate Administrator for Exploration, Michael Griffin,[62] also provided direction to this team. A Mars Study Team began work in August 1992 at the Lunar and Planetary Institute in Houston to address "the whys of Mars exploration to provide the top-level requirements." The group developed what it called "design reference mission" roadmaps, initially in 1993 and 1994, then in 1997; and in 1998, the team issued an update.[63]

---

59. Thor Hogan, "Lessons Learned from the Space Exploration Initiative," *NASA History News and Notes* (November 2007): 4.

60. See, for example, authors' interview with Dr. Ed Weiler, 22 September 2005, pp. 5, 30, NASA HRC.

61. NASA has 10 Field Centers dispersed geographically across the country. Since NASA's inception in 1958, Center Directors have had varying formal reporting relationships with NASA Headquarters.

62. Griffin left NASA shortly after this time period and returned as the Administrator in April 2005.

63. Michael B. Duke and Nancy Ann Budden, *Mars Exploration Study Workshop II: Report of a Workshop Sponsored by NASA Lyndon B. Johnson Space Center and Held at the Ames Research Center 24–25 May 1993* (NASA Conference Publication 3243, NASA JSC, 1993), available

The Mars Reference Mission examined sending a single mission of astronauts to Mars's surface with an implicit goal of capping total mission life-cycle costs at $20 billion. The initial version proposed a large (200-metric-ton) launch vehicle that would require four launches, but planners realized that a smaller (80-metric-ton) launcher would need only two additional launches, making that system more efficient.[64]

The Mars Reference Mission embraced several important goals, including seeking low-cost alternatives to finance human spaceflight, prioritizing some aspects of scientific discovery in mission design, and tightly integrating robotic and human exploration. The architecture of the Reference Mission built upon the Synthesis Group Report and Robert Zubrin's "live off the land" approach for in situ resource utilization, especially for heavy propellants. The reference

---

at *http://ntrs.nasa.gov/archive/nasa/casi.ntrs.nasa.gov/19940017410_1994017410.pdf* (accessed 18 March 2011). The preface of this report (p. iii) noted that the Associate Administrator for Exploration (Griffin) emphasized that the team should study the commonalities of lunar and Mars missions, but the team ended up choosing "an approach that emphasized the important aspects of Mars exploration without consideration of the lunar capability" because the former is "inherently more complex" than the latter. For two papers about the 1993 Mars Design Reference Mission, see David B. Weaver, Michael B. Duke, and Barney B. Roberts, "Mars Exploration Strategies: A Reference Design Mission" (IAF93-Q.1.383), *https://web.archive.org/web/20071218101048/http://ares.jsc.nasa.gov/HumanExplore/ Exploration/EXLibrary/DOCS/EIC044.html* (accessed 12 June 2018); and David B. Weaver and Michael B. Duke, "Mars Exploration Strategies: A Reference Program and Comparison of Alternative Architectures" (AIAA 93-4212), *https://web.archive.org/web/20070125043254/ http://ares.jsc.nasa.gov/HumanExplore/Exploration/EXLibrary/DOCS/EIC043.html* (accessed 12 June 2018). For the more well-known 1997 version, see Stephen J. Hoffman and David I. Kaplan, eds., *Human Exploration of Mars: The Reference Mission of the NASA Mars Exploration Study Team* (Houston, TX: Johnson Space Center NASA SP-6107, July 1997), especially pp. 1–5, available at *https://ntrs.nasa.gov/archive/nasa/casi.ntrs.nasa.gov/19980037039.pdf* (accessed 16 August 2018). For the 1998 addendum, see Bret G. Drake, ed., *Reference Mission Version 3.0: Addendum to the Human Exploration of Mars: The Reference Mission of the NASA Mars Exploration Study Team* (Washington, DC: NASA SP-6107, EX13-98-036, June 1998), available at *https://ntrs.nasa.gov/archive/nasa/casi.ntrs.nasa.gov/19980218778.pdf* (accessed 16 August 2018).

64. See Bret G. Drake, ed., *Reference Mission Version 3.0, Addendum to the Human Exploration of Mars: The Reference Mission of the NASA Mars Exploration Study Team* (Washington, DC: NASA SP-6107-ADD, June 1998), pp. 2, 9. Also see anonymous "Previous Exploration Studies" document, p. 8, in the Space Architect Doc Archive, CY00 DPT Phase 2, Historical Perspectives files, NASA HRC. Additional summary information about the Reference Missions is in a similar document by Douglas Cooke, "An Overview of Recent Coordinated Human Exploration Studies," January 2000, in the Space Architect Doc Archive, CY00 DPT Phase 2, Historical Perspectives files, NASA HRC. The follow-on to DPT, NEXT, looked carefully at the Reference Missions and took time to prepare a lengthy document summarizing them; see NEXT DRM Summary document in Space Architect, Doc Archive, CY02 NEXT, ACT files, NASA HRC.

mission was not meant to be "implementable in its present form"; rather, it would outline general concepts for future human exploration.[65]

The Mars Reference Mission team examined building infrastructure and technical expertise by sending astronauts to the Moon before going to Mars but decided to focus on Mars exploration without assuming that lunar capabilities would be a necessary precursor step. Repeating the exercise of searching for a compelling rationale for sending astronauts to Mars that many blue-ribbon commissions had done previously, the Mars Reference Mission team reached beyond the overused themes of inspiration and technological investment. The team contended that because Mars is the planet most like Earth, we should explore it to understand our own planet more thoroughly and to prepare eventually to move to Mars. The team also called for a large international cooperative component to a Mars mission.[66]

In terms of programmatic objectives, the Mars Reference Mission aimed to provide an alternative to a 30-year program costing hundreds of billions of dollars. Some of its distinguishing features included the following:

- No long-term LEO assembly sequences and no Mars orbit rendezvous before landing.
- Relatively short trips to and from Mars but long surface stays.
- A new heavy-lift launch vehicle that could send both crews and cargo directly to Mars with a minimum number of launches.
- In situ resource utilization on Mars.
- A common habitat module for both transit to and living on the surface of Mars.
- Bypassing the Moon.
- The possibility of aborting missions to the Martian surface.

The Mars Reference Mission aimed to accomplish these goals, all without making the risks "either more or less than other 'flags and footprints' scenarios for the first mission."[67]

The Moon also gained attention as a potential destination in the 1990s. In September 1995, Administrator Goldin initiated a study dubbed Human Lunar Return (HLR) to look at inexpensive ways to send astronauts back to the Moon utilizing existing technologies and in situ lunar resources. Run mainly out of JSC and led by Elric McHenry, HLR was an outgrowth of Goldin's "faster, better, cheaper" management approach. The study's ambitious goals were to put humans back on the Moon by 2001, expand human presence into space, and

---

65. Hoffman and Kaplan, *Human Exploration of Mars*, pp. III, IV, and p. 1–3.
66. Weaver, Duke, and Roberts, "Mars Exploration Strategies."
67. Ibid. The quotation is from the very end of this document.

eventually go to Mars. The Moon would be a "technology test bed" for sending humans to Mars and would use existing Shuttle and ISS hardware and personnel as much as possible. Perhaps ironically, one of the HLR "architectures" was similar to that of SEI, which was widely criticized as being much too expensive. The final HLR Report was issued on 7 August 1996. However, it was overshadowed by an announcement that same day from another JSC team. HLR ended up being short-lived and was "shelved" in late 1996.[68]

On 7 August 1996, NASA held a press conference to announce that a team of scientists from Johnson Space Center had uncovered "exciting, even compelling, but not conclusive" evidence of prior life on Mars.[69] David McKay's team had observed that microscopic shapes in a particular meteorite known to be from Mars appeared similar to tiny fossils left by Earthly bacteria. The rock, known as ALH84001, had been found in 1984 in the Allen Hills region of Antarctica. Planetary geologists and geochemists earlier determined that the rock had been ejected from Mars billions of years ago. What was in question was whether or not the rock contained definitive evidence of ancient Martian life, even at the lowly bacterial level.

The issue immediately came to the public's attention. President Bill Clinton made brief public comments expressing wonderment and asked NASA to investigate it further. He directed Vice President Al Gore to convene a "space summit" and used the opportunity to remind the American public of NASA's program of sending robotic spacecraft to Mars.[70] On 11 December 1996, Vice President Gore chaired an afternoon-long meeting of free-ranging intellectual debate with national experts on the scientific and social implications if ALH84001 proved conclusively that extraterrestrial life did exist. The next day, Gore issued

---

68. "Status of Human Lunar Return Study to JSC Center Director," 7 August 1996; and Marcus Lindroos, "Lunar Base Studies in the 1990s, 1996: Human Lunar Return (HLR)," available at *http://www.nss.org/settlement/moon/HLR.html* (accessed 26 October 2011).

69. The quotation is from "Statement from Daniel S. Goldin, NASA Administrator," NASA News Release 96-159, 6 August 1996, available at *http://www.nasa.gov/home/hqnews/1996/96-159. txt* (accessed 3 April 2008), which briefly described the discovery and announced that there would be a press conference the next day. Another press release was issued on 7 August; see "Meteorite Yields Evidence of Primitive Life on Early Mars," NASA News Release 96-160, *http://www.nasa.gov/home/hqnews/1996/96-160.txt* (accessed 3 April 2008). The McKay team's work was published as David S. McKay, Everett K. Gibson, et al., "Search for Past Life on Mars: Possible Relic Biogenic Activity in Martian Meteorite ALH84001," *Science* 273 (16 August 1996): 924–930.

70. William J. Clinton, "NASA Discovery of Possible Life on Mars," in "Remarks on Departure for San Jose, California, and an Exchange with Reporters," 7 August 1996, available at *http:// www.presidency.ucsb.edu/ws/print.php?pid=53170* (accessed 3 April 2008).

a statement supporting NASA's robotic space science program and, in particular, its Origins program, but he was not involved in the matter further.[71]

The issue faded over time as it became clear that McKay's JSC team could not prove its claims conclusively. However, the ALH84001 episode clearly reinvigorated interest in space exploration and Martian exploration in particular, whether robotic or human. The Mars Reference Mission issued in 1997 and 1998 gained special currency as a result of this episode. The attention also increased interest in NASA's nascent Origins program and the fledgling discipline of astrobiology.[72]

Also in 1996, aerospace engineer Robert Zubrin published his book *The Case for Mars*, in which he proposed a "Mars Direct" approach to living off the land (using supposed Martian physical resources to create oxygen to breathe and rocket fuel to return astronauts to Earth). A deliberately bare-bones alternative to what Zubrin perceived as NASA's overly complex and expensive approach to human spaceflight, *The Case for Mars* inspired the founding of the Mars Society and invigorated human exploration proponents.[73]

Inspirational events and pounds of studies, nonetheless, did not in themselves provide a foundation for moving beyond the achievements of the past. Administrator Daniel Goldin was one of a only few individuals at NASA with both the means and motivation to significantly influence the direction of the Agency. Shortly after his arrival at NASA in April 1992, Goldin made good on his reputation as an agent of change by instituting a management approach known simply as "faster, better, cheaper" (FBC).[74] Under the FBC approach, small, agile teams would design, build, launch, and successfully operate a new

---

71. For more on the "Mars rock" episode, see Kathy Sawyer, *The Rock from Mars: A True Detective Story on Two Planets* (New York: Random House, 2006); "Statement of Vice President's Space Science Symposium, 12 December," copy in file 9009, NASA HRC; and "Evidence of Ancient Martian Life in Meteorite ALH84001?" *https://planetaryprotection.nasa.gov/summary/alh84001* (accessed 10 July 2018).

72. Steven J. Dick, *Life on Other Worlds: The 20th Century Extraterrestrial Life Debate* (Cambridge: Cambridge University Press, 1998), pp. 68, 140; and "The Mars Rock," chap. 8 in *The Living Universe: NASA and the Development of Astrobiology*, ed. Steven J. Dick and James E. Strick (New Brunswick, NJ: Rutgers University Press, 2004), pp. 179–201.

73. Robert Zubrin with Richard Wagner, *The Case for Mars: The Plan To Settle the Red Planet and Why We Must* (New York: Free Press, 1996). Also see the Mars Society founding document at *https://web.archive.org/web/20080207141711/http://www.marssociety.org/portal/groups/tmssc/founding_declaration* (accessed 14 June 2018); and Robert Zubrin, "Pushing Human Frontiers" in *Looking Backward, Looking Forward: Forty Years of U.S. Human Spaceflight Symposium*, ed. Stephen J. Garber (Washington, DC: NASA SP-2002-4107, 2002), pp. 137–147.

74. Goldin gave a speech in May 1992 in which he used this terminology; see record number 31887, NASA HRC. File 15669 in the HRC covers Goldin and FBC and includes material starting in 1992.

generation of lighter, simpler spacecraft on compressed schedules. Having worked for several decades as a government contractor on classified space programs at TRW, Goldin was familiar with the aerospace community's metaphor that schedule, performance, and cost were three intrinsically linked sides of a triangle: managers could pick two of these three parameters, but the third one would inevitably suffer. He was also familiar with the aerospace community's common lament about the need to reduce launch vehicle costs for boosting payloads into space. Despite these presumed limitations, Goldin believed that improvements were possible and necessary. He decried the large, costly, and overly complicated scientific spacecraft that had been characteristic of NASA for many years. The 1993 loss of the Mars Observer provided further evidence of the wisdom of changing how NASA and its contractors made space vehicles.[75]

While not a product of the FBC initiative, the successful Hubble Space Telescope servicing missions increased confidence in the technical prowess of NASA engineers and provided evidence to NASA scientists that they too might have a compelling reason to support human spaceflight. Competition for resources within NASA often divided the human spaceflight and scientific communities from one another. The sense of division between the spaceflight and science communities reinforced the unfortunate but widely held notion that the two realms of activity had little to offer one another.

The science community learned otherwise with the experience of the HST, which NASA launched in April 1990 aboard the Shuttle. The first images sent back to Earth provided evidence that something was wrong with the first of NASA's space-based "Great Observatories." A NASA commission concluded that HST's main mirror was not the precise shape it was required to be to meet its scientific goals and objectives. The culprit was a miscalibrated instrument that caused workers to grind certain parts of the mirror slightly too much. This "spherical aberration" garnered much negative press attention. However, because the telescope was designed to be serviced by astronauts periodically, possibilities existed for fixing the problem. HST project engineers developed a corrective optics package that Shuttle astronauts installed on the first scheduled servicing mission in December 1993. The solution worked perfectly and extended the capacity of HST to do new and unique science. Leaders of NASA's scientific community learned an important lesson. The successful servicing mission demonstrated that human spaceflight and science were not necessarily in opposition

---

75. Howard E. McCurdy, *Faster, Better, Cheaper: Low-Cost Innovation in the U.S. Space Program*, New Series in NASA History (Baltimore: Johns Hopkins University, 2001), pp. 2–7 and passim.

and, indeed, scientific advancements in space could potentially depend on human spaceflight capabilities.[76]

The FBC approach was highly successful initially; only one of 10 spacecraft projects started between 1992 and 1998 failed.[77] Among the most notable early successes, the landing of the Mars Pathfinder spacecraft and the deployment of its robotic rover (Sojourner) on 4 July 1997 further fueled interest in Martian exploration. Landing on the Red Planet on the American Independence Day, it was one of the first major projects that NASA covered extensively on its Web site in "real time." Multiple "mirrored" servers were set up to handle the large number of people who logged on to the Web to track the mission. The Sojourner rover captivated the public's attention in a way no other robotic spacecraft had before. The rover was named after Sojourner Truth, the pseudonym for an abolitionist and women's rights advocate who lived in the mid-19th century.[78]

The mission's price tag, $265 million, was several times less than that of the two Viking spacecraft missions to Mars in the 1970s, providing a clear case for the benefits of the faster, better, cheaper approach.[79] The Mars Pathfinder was also an engineering and scientific success. Designed to be a technology

---

76. For general historical information about HST, see, for example, *http://history.nasa.gov/hubble/* and associated subpages (accessed 9 April 2008). Two notable histories of the HST are Robert W. Smith, *The Space Telescope: A Study of NASA, Science, Technology, and Politics* (Cambridge: Cambridge University Press, 1989); and Robert Zimmerman, *The Universe in a Mirror: the Saga of the Hubble Space Telescope and the Visionaries Who Built It* (Princeton, NJ: Princeton University Press, 2008). For more details about the spherical aberration, see, for example, Robert Capers and Eric Lipton, "The Looking Glass: How a Flaw Reflect Cracks in Space Science," *Hartford Courant* (March–April 1991). This series of four articles won a Pulitzer Prize.

77. McCurdy, *Faster, Better, Cheaper*, p. 2.

78. For more on the public's interest and specifically how it was such a major Web event, see Brian Dunbar, "Pathfinder Gets Hit Hard on the Internet," NASA Internet Advisory I97-8, 9 July 1997, and assorted news articles from that day, cited in the 9 July 1997 entry (pp. 81–82) of Marieke Lewis and Ryan Swanson, *Aeronautics and Astronautics: A Chronology: 1996–2000* (Washington, DC: NASA SP-2009-4030, 2009), available at *http://history.nasa.gov/sp4030.pdf* (accessed 18 March 2011). See *http://mars.jpl.nasa.gov/MPF/rover/name.html* (accessed 17 April 2008) for more information about the rover's name.

79. The first of NASA's large and expensive robotic planetary spacecraft, Viking was considered a success despite its high costs. The goals of the Viking project were to obtain high-resolution imagery of the Red Planet's surface, to analyze the structure and composition of the Martian atmosphere and surface, and to search for evidence of life. The Viking spacecraft transmitted scientific data to Earth for over seven years, far exceeding its designed three months of surface operations, although the spacecraft disappointed some critics because its instruments were unable to find or prove definitively the existence of life on Mars. David Williams, author/curator, "Viking Mission to Mars," *http://nssdc.gsfc.nasa.gov/planetary/viking.html* (last updated 18 December 2006) (accessed 21 July 2008); Launius and McCurdy, *Robots in Space*, p. 159; and McCurdy, *Faster, Better, Cheaper*, pp. 62–64. For more on Viking, see Ezell and Ezell, *On Mars*, passim and p. ix.

demonstrator for a subsequent Mars lander and rover, it utilized a tailored entry-descent-landing parachute and novel system of specially designed airbags so that the rover could land and then right itself and roll out onto the Martian surface no matter which way it touched down. This engineering system worked well, and the rover far outlived its predicted useful life (30 Mars days). Pathfinder's lander and rover also returned a great deal of scientific data, providing strong evidence that Mars once had a considerable amount of liquid water.[80] All of these factors combined to rekindle interest in further robotic exploration of Mars.

The reputation of the FBC approach remained untainted between 1992 and 1999, a period in which 1 of 10 spacecraft projects failed. A series of high-profile failures in 1999, however, undermined Goldin's signature initiative. In March of that year, the Wide-Field Infrared Explorer (WIRE) astronomy spacecraft failed. In September, NASA's Mars Climate Orbiter was lost after contractor mission designers and operators embarrassingly confused metric and imperial units of measurement. Then in December, the Mars Polar Lander spacecraft and its twin Deep Space 2 microprobes were lost upon atmospheric entry at Mars. With this string of failures, engineers and managers in the space community began to question the wisdom of faster, better, cheaper.[81]

While Administrator Goldin then scaled back his FBC ambitions for robotic scientific spacecraft, he and other advocates did not lose sight of the ultimate FBC goal: a less expensive human mission to Mars. Some of its advocates believed that the FBC approach could still provide such great efficiencies that astronauts could go to Mars for one-third the cost of the Apollo program. By one estimate, adjusting for inflation, this would be about $40 billion in fiscal year (FY) 2000 dollars. This would be an order of magnitude smaller than SEI's estimated price tag of $400–500 billion. Goldin challenged planners at JSC essentially to adopt Zubrin's minimalist approach to cut potential costs for human expeditions to Mars dramatically.[82] While the effort did not yield many tangible results (some might say it was doomed not to), these early FBC

---

80. See *http://www.nasa.gov/mission_pages/mars-pathfinder/* (accessed 17 April 2008).

81. McCurdy, *Faster, Better, Cheaper*, pp. 2–7 and passim.

82. McCurdy, *Faster, Better, Cheaper*, pp. 148–149. McCurdy uses the figure of $21.3 billion in real-year dollars for the Apollo program through its first landing on the Moon in 1969. Adjusted for inflation, this would be approximately $120 billion in year 2000 dollars. One-third of that total would be $40 billion. *Faster, Better, Cheaper* was published in 2001, so it adjusted for inflation to FY 2000 dollars. Since the peak year for Apollo funding was 1965, adjusting $21.3 billion in then-year dollars for inflation would yield approximately $207 billion in FY 2018 dollars. This calculation was done via the NASA New Start Inflation tool at *https://web.archive.org/web/20120902044105/http://cost.jsc.nasa.gov/inflation/nasa/inflateNASA.html* (accessed 7 November 2018).

experiments inspired confidence within the space community that effective managers could improve the cost and speed of spacecraft development projects.

On the Shoulders of Giants

By the end of the 1990s, a rich body of ideas, as well as a dramatic history of success and failure, lay available to consult for government leaders interested in crafting a new plan for the nation's space program. The store of ideas included a dominant paradigm that focused heavily on human spaceflight with no significant scientific component; a competing vision that recommended a robotic-only, science-driven program without the drama of human spaceflight; and a range of alternative visions in between. All drew selectively on deeper bodies of knowledge—fictional, speculative, and technical—that could help bridge the gap between the present and the distant future. In this world, slingatrons, ballistic cannons, blast wave accelerators, and large coilguns appeared as exciting possibilities for decreasing the cost of access to space rather than as visions of a mad scientist or the off-duty preoccupation of ambitious engineers.

The reports of formal commissions, Agency working groups, and postmortem "lessons learned" studies provided rationales, strategies, and concepts for the future of the space program. The Ride Report in 1987 recommended a broad space program that relied on both robotic missions throughout the solar system and human missions to Mars and the Moon to further the cause of exploration and achieve scientific goals. In addition to advocating for a flexible approach to exploration based on available resources, the go-as-you-pay approach, the 1990 Augustine Commission reiterated several proposals of the Ride Report, including devising shared goals for human and robotic spaceflight, improving NASA's Earth observation capabilities, and sending humans to Mars. The 1991 Stafford Commission added flesh to earlier concepts, devising four elaborate space architectures as potential options for guiding the human spaceflight program. Other less visible study groups, such as the Mars Reference Mission and Human Lunar Return, updated old ideas and proposed novel concepts for future missions.

After the demise of SEI, NASA officials were reluctant to embark on a high-profile exercise in planning for human missions beyond LEO. Nevertheless, increased public interest in space following several notable achievements—such as the Mars Pathfinder Mars landing in July 1997, the repair of the Hubble telescope, and the discovery of possible life on an old Martian rock in Antarctica—raised hopes within the Agency that future mission plans might include a human visit to Mars. The cautious approach to planning that NASA would adopt reflected both the peculiar character of the Agency's leadership and a larger political context unfavorable to bold initiatives in space.

3

"SNEAKING UP ON MARS": ORIGINS OF THE DECADAL PLANNING TEAM

THE HISTORY of the Decadal Planning Team begins in late 1998, when OMB provided funding for NASA to initiate a new long-term planning effort and NASA Administrator Daniel Goldin embraced this idea. This chapter explores the first three years of DPT, focusing on the rationale for establishing it and the people, events, and ideas that defined the team's early history. The chapter explains how the team functioned, the principles that guided it, and the obstacles it faced in meeting the expectations of OMB and NASA's leadership from 1998 to Administrator Daniel Goldin's departure from NASA in November 2001.

Seed Money, Seed Ideas

In the fall of 1998, the Office of Management and Budget inserted language into NASA's fiscal year 2000 budget request during the "passback" stage of the budget review process.[1] Under the heading "Next Decade Planning," OMB directed NASA to add $5 million to its budget requests for a new planning initiative "to explore and refine concepts and technologies" that would drive the Agency's

---

1. OMB works on behalf of the White House to prepare the budgets of individual federal agencies for inclusion in the President's annual budget request, which the White House typically submits to Congress in February each year. Prior to that, after reviewing and modifying agency budget requests, OMB sends the budget documents back to the individual agency for further review, typically in November or December, before submission to Congress. This is known as the "passback." OMB often inserts programmatic language or funding in agency budgets, as occurred in the instance discussed above.

agenda through the next decade.[2] Both OMB and the Office of Science and Technology Policy (OSTP) expected the planning effort to generate "a varied menu of discrete options with viable plans along a spectrum of investment levels that would achieve well-defined goals before 2010."[3] The convoluted language of the document masked the decisive intentions of the OMB budget officers who inserted the wording. The budget officers were seeking concrete ideas for NASA's mission after the completion of the International Space Station. OMB expected NASA to use this funding to investigate ideas for reducing launch costs, improving the Agency's public outreach, expanding efforts to capture imagery of extrasolar planets, and establishing a permanent robotic presence in space for the purpose of conducting scientific research. Most significantly, OMB asked NASA to consider options for moving human spaceflight activities beyond LEO.[4] Budget officers at OMB with significant responsibility essentially directed NASA to investigate the possibilities for going where the Agency had dared not venture for nearly 30 years prior to that point.

Steve Isakowitz, who was then the Chief of Science and Space Programs at OMB, promoted this decadal planning initiative despite skepticism from "some very senior people at OMB and elsewhere in the White House [who thought that] the idea of getting NASA to think about going beyond low-Earth orbit when they can't even master building something in low-Earth orbit with the Space Station was heresy." These senior officials thought it would be unwise to add any funding to NASA's budget, even the relatively small amount of $5 million, for studying the possibilities for human spaceflight beyond LEO when the Agency was having financial problems with the ISS. Isakowitz reported that "a couple of people warned me not to do it [add the $5 million to NASA's budget]" because NASA officials might interpret it as a sign that the administration would be receptive to providing additional funding to support programs conceived with the seed money.[5]

---

2. "Next Decade Planning Funding History" document in Thronson budget materials, 15 September 2005, NASA HRC.

3. Congress established OSTP in 1976 and authorized it to take the lead on Government-wide science and technology activities and policies. OSTP's influence has varied in accordance with its leadership and each presidential administration's views toward science and technology. See *http://www.whitehouse.gov/ostp/about* (accessed 21 November 2017). For the quotation, see 1999 OMB Passback language, p. 14 (5 of 9 in fax), fax dated 19 December 1998, attached to DPT charter, NASA HRC.

4. OMB Passback language, p. 15 (6 of 9 in fax), fax dated 19 December 1998, attached to DPT charter.

5. Steve Isakowitz interview by Glen Asner and Stephen Garber, Washington, DC, 23 March 2006, pp. 7–8, 15, NASA HRC.

Isakowitz saw the situation differently. He agreed that difficulties control-ling costs and technical problems with the International Space Station and Space Shuttle Programs had damaged the Agency's reputation. He also knew that some policy-makers, including a few top officials in the White House, viewed the space agency as beleaguered and unfocused. NASA's lack of commu-nication with the White House, Isakowitz thought, made the situation worse. Yet he was well aware of the Agency's positive attributes, including its strong science and engineering capabilities and the great achievements the Agency had accomplished in the past. He also sensed that public sentiment about the Agency was more complicated than his colleagues thought. Opinions of the human spaceflight program at times seemed to range from indifference to con-tempt, but large numbers of Americans still looked to the Agency to accomplish feats otherwise considered impossible. NASA's robotic spacecraft, furthermore, received praise in the media and met with little political opposition. The budget passback language, in fact, praised NASA's "strong performance" in space sci-ence, specifically its success in implementing "faster, better, cheaper" missions such as those in the Discovery and New Millennium programs.[6]

Isakowitz believed that providing NASA with a relatively small amount of money to reconsider its mission could yield big dividends. As he later explained, "if you could put a stock value on the human spaceflight program, it was prob-ably at its lowest point in the early '90s and maybe that's why it seemed like such a good time to put a little bit of money aside" to invest in NASA.[7] Isakowitz thought that NASA could rebuild its reputation and its core skills base gradu-ally with some small, near-term successes. A few concrete successes would bol-ster the Agency's reputation in the eyes of lawmakers and the public and thereby

---

6. The Discovery program sought proposals for an entire space science mission coordinated by the Principal Investigator. The missions were designed to launch within 36 months of the proj-ect start, and the missions were cost-capped (initially at under $299 million, which was raised to $360 million and, starting in 2006, $425 million). The Near Earth Asteroid Rendezvous (NEAR) mission was the first Discovery spacecraft to be launched. See *https://science.nasa.gov/solar-system/programs/discovery* and *https://solarsystem.nasa.gov/missions/near-shoemaker/in-depth/* (accessed 8 June 2018), and Jefferson Morris, "NASA To Release New AO for Discovery Program on Jan. 3," *Aerospace Daily* (23 December 2005) for more details. The New Millennium Program was begun jointly by the Offices of Space Science and Earth Science in 1995 to develop advanced technologies for robotic spaceflight, with scientific data sometimes being an ancillary benefit. An example was the Deep Space 1 mission that tested ion propul-sion technology. See *http://nmp.jpl.nasa.gov/TECHNOLOGY/innovative-tech.html* (accessed 26 October 2011) for more details. The OMB discussion is from the 1999 OMB Passback language, pp. 11 (2 of 9 in fax) and 14 (5 of 9 in fax), fax dated 19 December 1998, attached to DPT charter. For more on Goldin's "faster, better, cheaper" approach, see McCurdy, *Faster, Better, Cheaper*.

7. Isakowitz interview, p. 15.

pave the way for a future president to endorse NASA's aspirations to take astronauts beyond LEO.

Isakowitz was not alone in believing that the time had come for a change in direction. Independently, NASA Administrator Daniel Goldin had reached the same conclusion. With the 1998 midterm elections complete and the Clinton administration moving into its final years, Goldin concluded that it was time to start putting together policy proposals for the next president. Other than initiatives to strengthen international cooperation, the Clinton administration instituted no major changes in space policy and produced few formal civilian space pol-

NASA's longest-serving Administrator, Daniel S. Goldin. (NASA)

icy documents. One well-placed policy-maker, John Schumacher, recalled that the White House thought NASA had its hands full managing current programs and merely wanted the Agency to focus on completing the International Space Station within budget and operating the Space Shuttle safely.[8] Sending humans beyond LEO was not a serious consideration at the time, yet Goldin sensed that a change in administrations might provide an opportunity for NASA to strike out in a bold new direction. Should the next presidential administration prove more responsive to human spaceflight, Goldin wanted to be ready with a well-thought-out plan that reflected the best ideas the Agency could offer.

8. See Schumacher interview Asner and Garber, NASA Headquarters, Washington, DC, 26 January 2006, p. 4; and Alan Ladwig's personal notes from 24 March 1999, which reflect more explicit, if still informal, guidance from the Clinton administration on space policy (both in NASA HRC). Ladwig was a senior advisor to Goldin who had served previously as the Associate Administrator for Policy and Plans. Assuming NASA could work within its existing human spaceflight budget, the Clinton administration was amenable to an incrementally building program that allowed flights to multiple space destinations. Frustration and disappointment both within and outside NASA were mounting over the future direction of human spaceflight, as repeated redesigns and the related delays and cost overruns on the development of the ISS meant that its first crew had not been launched yet (Expedition 1 began its mission on 31 October 2000). For some indirect language on this point, see section 3a of the Civil Space Guidelines section of the National Space Policy, 19 September 1996, at *http://history. nasa.gov/appf2.pdf*.

Key events that raised the public profile of NASA also made the late 1990s appear to be a good time to rethink the Agency's future. In 1998, the general public expressed great interest in the STS-95 Space Shuttle mission, which returned Senator John Glenn to space over 30 years after his historic Mercury mission. On the robotic space science side, the Mars Pathfinder rover that landed on 4 July 1997 was wildly popular and also contributed significantly to scientific understanding of the Martian surface. In addition, the 1996 discovery of a Martian rock in Antarctica that may have contained a microfossil of a living organism stirred great interest in astrobiology as a discipline that could profoundly impact average citizens by answering the fundamental question, Are we alone in the universe? These two Mars robotic successes led to the establishment of NASA's Origins science program, which in turn led to a variety of significant, new space science programs, including a follow-on to the Hubble Space Telescope, which became known as the James Webb Space Telescope.[9]

As an indication of his enthusiasm for refocusing the Agency, Administrator Goldin activated his new decadal planning working group in the spring of 1999, even though the formal funding for the group's activities that OMB had proposed and congressional appropriators had approved would not be available until October of that year.

On Sunday, 4 April 1999, Goldin held a meeting at his home on Capitol Hill with a small handful of top Agency officials to initiate a new long-range planning effort.[10] Goldin, who had a reputation for secrecy, instructed the officials at the meeting to refrain from mentioning anything about it to anyone not in attendance. He held this small initial discussion outside NASA Headquarters

---

9. Wesley Huntress discussed these events and the ascendancy of NASA's Origins in a speech he gave to the American Astronautical Society in Houston, TX, on 17 November 1998 titled "Grand Challenges for Space Exploration." In his speech, Huntress mentioned that Carl Sagan had sent him a letter noting some of these key events and how the time was propitious to redirect NASA. This "Grand Challenges" speech was in turn a formative element for DPT's early architects, as will be discussed later in this chapter. See "Huntress, 'Grand Challenges'" folder in DPT/VSE Assorted Background files, NASA HRC.

10. Harley Thronson, conversation with Steve Dick, 2 November 2004, transcript in the NASA HRC. Joe Rothenberg recalls that just he and Weiler attended this meeting, although perhaps a few others were there as well. See email from Joe Rothenberg to Steve Garber and Glen Asner, 24 October 2008, and his oral history, 14 September 2005, p. 3, both in the NASA HRC. Ed Weiler recalls that in addition to himself, Rothenberg, and Goldin, Sam Venneri and possibly George Abbey attended. (See Weiler interview, 22 September 2005, p. 3, HRC). At that time, Venneri was the Agency's Chief Technologist and had a close working relationship with Goldin. Abbey, then the Johnson Space Center Director, was an influential insider who had served in significant positions at NASA Headquarters and JSC since the late 1960s. (See George Abbey biographical file, file number 2, NASA HRC).

to foster creative thinking, to emphasize the significance of the effort, and to convey the importance of keeping it secret.

In addition to sensing that a change of presidential administrations could provide an opportunity for a major policy shift, Goldin thought the time was right from an economic perspective. The defense drawdown following the end of the Cold War, combined with high tax revenues from the dot.com boom of the late 1990s, allowed the U.S. government to run a budget surplus in 1998 for the first time in 29 years.[11] With federal economists projecting surpluses for a decade into the future, government and business leaders grew increasingly confident about the long-term economic health of the nation.[12] Administrator Goldin, in this context, entertained the possibility that NASA could get a significant increase in its budget. He thought it would be prudent for the Agency to be prepared with an ambitious, intellectually coherent new strategy should Congress or the White House offer additional money for new programs.[13]

Yet Goldin's decision to create the working group that became DPT and to put his strong support behind it went far beyond mere positioning to take advantage of the political and economic opportunities of the moment. Goldin was seeking to reshape the Agency for two primary reasons. First, he wanted to create a greater sense of unity across disparate organizational units and geographically dispersed Centers. Goldin believed that culture, geography, and habit resulted in inefficiencies and conflicts that were against NASA's long-term interests. Programs, projects, and entire Field Centers operating in isolation failed to share information and took actions without consulting one another. Second, Goldin sought to change the way NASA and its supporters understood the Agency's long-term mission and conveyed that mission to the public. The public deserved a solid practical or scientific justification for human exploration of Mars before the Agency asked citizens to commit their tax dollars.[14]

As with most large, dispersed organizations engaged in multiple, loosely related activities, NASA struggled throughout its history with coordination across its 10 Field Centers. The complexity of the organization, the dynamic

---

11. Congressional Budget Office, *The Budget and Economic Outlook: Fiscal Years 2008 to 2017*, Pub. No. 2941, January 2007, p. 140.

12. Congressional Budget Office, *The Economic and Budget Outlook: An Update*, 1 August 1998, p. 35.

13. Daniel Goldin, interview by Asner and Garber, George Washington University, Ashburn, VA, 16 March 2006, pp. 2–6, NASA HRC. Ed Weiler mentions this broader economic context in his 22 September 2005 oral history. See pp. 43–44.

14. Goldin interview, passim. Also see W. Henry Lambright, *Transforming Government: Dan Goldin and the Remaking of NASA* (Arlington, VA: PricewaterhouseCoopers Endowment for the Business of Government, March 2001).

nature of its activities, and regular changes in leadership meant that even the most well-thought-out efforts to stabilize organizational processes and reporting structures were fleeting. Despite the considerable difficulty of changing NASA's organizational dynamics, several Administrators have tried to impose order on NASA's disparate facilities and activities. Improving the effectiveness of the Agency as a whole without harming individual programs, projects, or Field Centers was not an easy task. The challenge was to provide enough independence to individual units so that they could pursue their goals with the fewest possible constraints while at the same time ensuring that each Center did its part to maximize the effectiveness of the entire Agency and not hinder other Centers or programs.[15]

A significant schism divided the Agency's robotic space science and human spaceflight activities at both the Headquarters and Center levels. Competition for resources, prestige, and power, as well as substantial cultural differences, account for the schism historically. On the one hand, scientists in various robotic disciplines such as Earth science, astronomy, and planetary exploration often have viewed the human spaceflight program, by far the largest and most costly of NASA's programs, as lacking a tangible purpose (beyond its considerable inspirational value) and thus a drain on the Agency's resources. On the other side, those engaged in human spaceflight tended to view their activities as the core reason for the Agency's existence and emphasized the enormous responsibility involved in assuring the safety of human spaceflights.[16] The two points of view could be summarized as either a budgetary zero-sum game or a "rising tide lifting all boats."

Goldin believed that the two major elements of the Agency had much to offer each other, despite their differences. The Office of Space Science's (OSS's) strategic planning process, most notably, set a strong example within NASA. In addition, the robotic space science community set its priorities through the National Academy of Sciences' decadal surveys (separate from what became known as NASA's Decadal Planning Team) for various fields of astronomy; no comparable surveys were produced for human spaceflight. Goldin wanted to apply the robotic science community's planning approach to human exploration. As Goldin reflected afterward, space scientists were not inherently better long-range planners than their human spaceflight counterparts; rather, the leaders of the Office of Space Flight rarely had time for reflection on NASA's long-term plans because they were tightly focused on their immediate responsibilities, which included managing large, complex, operational programs, such as

---

15. McCurdy, *Inside NASA.*
16. Goldin interview, passim.

the International Space Station, and ensuring the immediate health and safety of people in space.[17]

Yet it was not just about the superior long-term planning abilities of the Office of Space Science. Goldin wanted an intellectually compelling rationale to underpin NASA's potentially grand plans. He believed that science itself had something to offer human spaceflight. Despite claims about the technological and economic spinoffs from the human spaceflight program, some space scientists contended that the increase of human knowledge based on decades of human spaceflight was meager, with the exception of studies of the properties of the Moon and the impact of the space environment on human physiology.[18] Goldin also objected viscerally when he saw Mars enthusiasts wearing "Mars or bust!" buttons because he felt that this sentiment overlooked the larger, fundamental reasons to send humans to Mars. Beyond the inherent allure of human spaceflight to a limited portion of the American population, why should the government fund such an endeavor? As Goldin expressed later, a "*Field of Dreams*; build it and they will come" mentality was insufficient for a major program that needed to engage the nation.[19] Scientific discovery was the most intellectually justifiable and compelling rationale Goldin could conceive. Overcoming organizational divides with genuine cooperation between the two major halves of the Agency, Goldin believed, would allow NASA to begin moving toward a truly unified space program that exploited the capabilities of robots and humans to achieve the goal of expanding scientific knowledge.

Goldin met at Headquarters with some of his top aides on 15 April 1999 and then again on 19 April to formulate more of the parameters for what they

---

17.   Goldin interview, pp. 11–12, 16–17. In terms of OSS's leadership in long-term planning, Mark Saunders observed that "Code S has done a masterful job of origin, evolution, and destiny" and suggested that DPT should expand on this theme. (NASA has traditionally been organized into "Codes," similar to departments or other high-level organizational units, and Code S was the Office of Space Science.) Also see appendix E of this book. See "First Meeting of the Decadal Planning Team for Exploration" (hereafter "First DPT meeting"), vol. II (25 June 1999), transcript, pp. 220, 292, \Nguyen\zip2\minutes subfolder, DPT files, NASA HRC. Dennis Bushnell also observed that OSS had a strong advanced technology program, but that NASA's Human Exploration and Development of Space organization was very poor in this regard. See First DPT meeting, vol. II (25 June 1999), transcript, p. 108.

18.   There have been numerous studies, many of which lack economic rigor, that have attempted to quantify the return on investment in the aerospace sector. Most observers conclude simply that research and development (R&D) investment pays off, even if it is hard to quantify. One concise article on this subject concludes by quoting a space economist as saying that for a nonprofit government organization such as NASA, "any ratio of economic benefits versus spending that exceeds 1-1 'is a success.'" See Ken Chamberlain, "Measuring the NASA Stimulus: Estimates Vary Widely on How Much the Space Agency's Spending May Benefit the Economy," *National Journal* (27 August 2010).

19.   Goldin interview, p. 3.

already referred to as "decadal planning." It is not known who attended each meeting, but apparently some staff from OMB (including presumably Steve Isakowitz) attended the 19 April meeting. They talked about a broad strategy that would be informed by previous studies about the future of human and robotic spaceflight. They also discussed the possibility of using savings from the recently consolidated Space Shuttle operations contract to pay for a much grander project.[20]

Goldin oversaw another significant "interenterprise working group" meeting on 10 May, off-site in Washington, DC. The Administrator explained to the larger group that he wanted a "virtual Center" team set up to take a fresh look at an integrated programmatic strategy for the whole Agency, in time for the new presidential administration that would take office in 2001. As Goldin put it later, "the new president and his team would have their own ideas, but I felt NASA owed whoever was going to be president, unbiased and factual information."[21] A NASA Human-Robotic Exploration Team (HRET), based out of Johnson Space Center, was already engaged in long-range planning, but Goldin indicated that he wanted less expensive options. In general, Goldin was frustrated by what he saw as JSC's large, complex, and costly approach to human exploration. He pushed for the adoption of a more incremental approach that would cost less and generate little political opposition. Goldin specifically called for a new team to "develop a range of concepts which address a long[-]term integration of robotic and human exploration objectives" as well as "scientific and human infrastructure." He took a personal interest in this new team and joked that he wanted to be its project manager.[22]

At this May meeting, Goldin tasked Joe Rothenberg and Ed Weiler, the heads of human spaceflight and space science at NASA, respectively, to select a team and its leader and for Rothenberg to report back to him. Goldin asked Sam Venneri, NASA's Chief Technologist, to start on a draft of the charter for the group. Dan Mulville, NASA's Chief Engineer, worked with Alan Ladwig, a senior advisor to Goldin who had served previously as the Associate Administrator for Policy and Plans, and Lori Garver, then the Associate Administrator for Policy and Plans, in drafting the terms of reference.[23] After the initial effort by Sam Venneri, the charter was completed by an ad hoc "executive

---

20. Alan Ladwig personal notes, 1, 15, and 19 April 1999.
21. Goldin interview, p. 4.
22. Alan Ladwig personal notes, 10 May 1999.
23. Alan Ladwig personal notes, 10 May 1999 and 1 June 1999; Dan Mulville to Steve Garber, 2 December 2005, Printed Emails to Glen Asner and Steve Garber, NASA HRC.

Ed Weiler, Associate Administrator for Space Science. (NASA Weiler-ofc2001)

Joe Rothenberg, Associate Administrator for Spaceflight. (NASA)

committee" made up of Jim Garvin, Lisa Guerra, and Harley Thronson, with significant input from Weiler and Rothenberg.[24]

June 1999: Phase 1 Starts with the Assembling of the Team

On 1 June 1999, Ed Weiler and Joe Rothenberg formally created DPT. Both men reported directly to NASA Administrator Dan Goldin, having assumed their jobs as Associate Administrators in 1998. Having worked together in other capacities for many years, Rothenberg and Weiler enjoyed a close professional and personal relationship. Most notably, they worked together on the first Hubble Space Telescope servicing mission in December 1993, when Weiler was the HST Project Scientist and Rothenberg was the Associate Director of Flight Projects for Hubble at NASA's Goddard Space Flight Center (GSFC). An astrophysicist by training, Weiler joined NASA in 1978, shortly after receiving his Ph.D. In 1979, he became the program scientist for HST, a position

---

24. See Alan Ladwig personal notes, 10 May 1999; Jim Garvin to Steve Garber and Glen Asner, 16 September 2005; and Joe Rothenberg to Steve Garber and Glen Asner, 24 October 2008, Printed Emails to Glen Asner and Steve Garber, NASA HRC.

Jim Garvin, DPT chair. (NASA)

he maintained for two decades while he also took on other increasingly important managerial positions. After temporarily serving as the Associate Administrator for the Office of Space Science, Weiler took this job permanently in November 1998. An engineer by training, Rothenberg became the Director of GSFC in 1995. He moved to NASA Headquarters in January 1998 to become the Associate Administrator for (Human) Space Flight.[25]

Al Diaz, the GSFC Director, and Goldin named Jim Garvin, a GSFC Earth and planetary geologist, as the DPT chairperson. Garvin's knowledge of the lunar, Martian, and Earth environments was broad and deep. He had worked both with NASA's scientific and human spaceflight communities, and had experience flying scientific payloads aboard the Shuttle. He believed that he was selected because he had worked on both human spaceflight and robotic spacecraft and also possessed an advanced understanding of the science of Mars and the Earth.[26] Weiler remembers that both he and Rothenberg recommended Garvin to Goldin.[27] His outgoing personality also set him apart.

Lisa Guerra was tapped to be the manager. General John R. "Jack" Dailey, then the NASA Deputy Administrator, had been talking with her about her working for him and then working on DPT. Guerra, a GSFC employee at the time, had worked on special assignments for Rothenberg during his tenure as Center Director. She was also familiar with Weiler because of her GSFC role as the Program Integration Manager for the Next Generation Space Telescope (now the James Webb Space Telescope) while Weiler was the Origins Theme Director. Her engineering background spanned work experience in both human spaceflight and robotic missions while working at NASA

25. Regarding the close personal ties between Rothenberg and Weiler, see, for example, Ed Weiler, oral history, 22 September 2005, pp. 5–6. For more information on Weiler, see his biographical file (009672) in the NASA HRC. For more information on Rothenberg, see his biographical file (16097) in the NASA HRC.

26. Jim Garvin, interview by Asner and Garber, NASA Headquarters, Washington, DC, 18 October 2005, p. 6, NASA HRC.

27. Weiler oral history, p. 24.

JSC, Headquarters, and GSFC. With respect to exploration, Guerra had worked directly on many of the previous efforts, including the 90-Day Study, the Synthesis Group, and the SEI Program Office's First Lunar Outpost. Her broad Agency experience, as well as familiarity with much of NASA's long-term human exploration initiatives, enabled DPT to get off to a quick start.[28] Guerra and Garvin had worked together before, on an Earth science effort, and had an effective collegial work relationship.

Rothenberg and Weiler asked the various Center Directors and leaders of Headquarters Enterprises (major

Lisa Guerra, DPT manager. (NASA)

programs) to nominate a capable person from each of their organizations who would have their leaders' full support.[29] Rothenberg and Weiler encouraged them to pick bold thinkers who also could work well as part of a team, sharing in the give-and-take of ideas. The Center Directors seemed to understand the importance of this team and hand-selected members (the team started with 20 members). (See appendix D-2.)

Goldin viewed DPT as different from many other such task forces and advisory planning boards, whose members traditionally have been established "greybeards" with broad reputations and experience. Instead, Goldin wanted fresh, youthful thinkers formulating long-term goals because they might still be working at NASA when those goals were realized.[30] The team's youth was

---

28. Lisa Guerra, oral history by Asner and Garber, NASA Headquarters, Washington, DC, 26 October 2005, pp. 2–3, 14–16, NASA HRC.

29. In 1999, the Directors for NASA's Field Centers primarily responsible for human space-flight (Johnson, Stennis, Marshall, and Kennedy) reported to the Associate Administrator for Space Flight (Rothenberg); the Director of the Jet Propulsion Laboratory reported to the Associate Administrator for Space Science (Weiler). Al Diaz, then the Director of Goddard, reported to the Associate Administrator for Earth Science (Ghassem Asrar). Space Flight, Space Science, and Earth Science were considered programmatic enterprises, along with Aero-Space Technology (headed by Spence Armstrong) and Life and Microgravity Sciences and Applications (headed by Arnauld Nicogossian). See *http://history.nasa.gov/orgcharts/orgchart6-99.pdf* (accessed 19 April 2018) for NASA's organizational structure at that time.

30. Goldin joked that he wanted all the team members to be under 40 years old. See Guerra oral history, 26 October 2005, pp. 5–6.

Harley Thronson, DPT member. (NASA)

exemplified by members in their mid-40s, such as Don Pettit, a scientist who had been selected as an astronaut in 1996 but had not flown in space yet, and Matt Golombek, a young JPL scientist whose profile had been raised by the success of his Mars Pathfinder team in the mid-1990s.[31]

Guerra had the opportunity to influence the selection of other team members. She floated the names of some JSC exploration colleagues, but these were rejected, evidently because they were considered to be overly committed to existing exploration approaches. She and Garvin both wanted Mark Saunders, a systems engineer then at Langley Research Center (LaRC) with experience working for the Navy on nuclear submarines and at NASA on a variety of space science missions. Saunders had joined NASA in 1989 to work on Space Station Freedom and had worked with Guerra when the two were involved in selecting the Discovery, Explorer, and Earth System Science missions.[32]

On 1 June, Rothenberg set the team in motion with an email informing the team members that they had the backing not only of the Center Directors, but also of the Agency's senior leadership (himself, Weiler, and Goldin). The team had 16 members, all but 2 of whom were from the Centers initially. In keeping with DPT's setup as an Agency-wide group, the team held its first meeting via

---

31. Born in 1955, Pettit earned a doctorate in chemical engineering and had worked at Los Alamos National Laboratory on microgravity science from 1984 to 1996. He was a member of the Synthesis Group blue-ribbon commission after the Space Exploration Initiative was announced in 1989 and also served on the 1993 Space Station Freedom Redesign Team. See *http://www.jsc.nasa.gov/Bios/htmlbios/pettit.html* (accessed 6 August 2008). Approximately the same age as Pettit, Golombek earned a Ph.D. in geology/geophysics. He was a planetary geologist specializing in Mars at JPL. In addition to being the high-profile Project Scientist for the Mars Pathfinder program since 1994 and thus partly responsible for a very successful "faster, better, cheaper" mission, Golombek was an active and prolific writer of scientific papers. See *http://mars.jpl.nasa.gov/MPF/bios/golombek.html*, *http://science.jpl.nasa.gov/people/Golombek/*, and *http://mars.jpl.nasa.gov/MPF/bios/golombek-bio.html* (all accessed 6 August 2008).

32. Guerra oral history, 26 October 2005, pp. 6–8, 15. For biographical information about Saunders, see files in the NASA HRC.

teleconference a week later, on June 8.[33] (See appendix E for more on NASA's organizational structure.) Headquarters enterprises soon contributed a few individuals to the team.

A few weeks later, DPT members (see appendix D-2) assembled in Washington, DC, on 24 June, for their first physical meeting—an intensive two days of presentations and discussions.[34] Garvin verbally summarized the ground rules for the team as he understood them, elaborating on Rothenberg's email. He told the group that their main task was to design a new joint robotic-human exploration strategy for NASA. The "critical constraints," as Garvin saw them, were that DPT should include astronauts where appropriate but not just consist of "flags and footprints" (some critics of human spaceflight saw the Apollo program as little more than this symbolism) and that DPT should be "science driven, technology enabled, affordable, and engaging."[35] All of those present likely had their own ideas as to what such terms meant, but the overall meaning for the group was not clear yet. What measures, for example, would the team use to gauge affordability? What strategy would their vision contain to ensure affordability? Whom did the team members seek to engage with their final product? Building a new Agency strategy meant translating concepts such as

---

33. Rothenberg wrote that DPT should take a "fresh look" at how to develop a "NASA Vision." He also noted that the "Blue Ribbon Team" would "work independent[ly] of Center Roles and responsibilities in what I would like to call a Virtual Center environment." Roger Crouch and Harley Thronson were the only Headquarters team members at that time, representing Codes M and S, respectively (copy of email from Joe Rothenberg, 1 June 1999, in binder 2, file 11, Lisa Guerra files, NASA HRC). See also Report of the NASA Exploration Team, "Setting a Course for Space Exploration in the 21st Century: June 1999 Through December 2000," NASA HRC.

34. The meeting took place on 24–25 June 1999. Verbatim transcripts of both daily sessions are available, as well as executive summaries. See First DPT meeting, vol. I and II (24 and 25 June 1999), transcript, NASA HRC. On day one, astronomer Harley Thronson presented on space science Grand Challenges and the "Living With a Star" Sun-Earth connection public outreach effort, physician David Dawson discussed medical and psychological issues, Dennis Andrucyk talked about the Earth science vision, Matt Golombek addressed robotic space science, and Lisa Guerra gave an overview of some recent strategic studies. Day two included other group exercises, selected presentations, and discussion.

35. Garvin used the terms "science driven," "technology enabled," "affordable," and "engaging" as bullet points under "critical constraints" in an outline he prepared for this meeting. Another bullet point was "humans in the loop where appropriate," and Garvin hand-wrote a note to himself under this bullet point that this was "not flags & footprints." See Jim Garvin, "First Meeting of Decadal Planning Team for Exploration," p. 4, in Early Brainstorming, APIO files, NASA HRC. While not coined in the DPT context, the disparaging "flags and footprints" term ended up being used frequently in discussions. Garvin used this term at the initial meeting several times, and at one point, David Dawson, an M.D. from Johnson Space Center, questioned whether this effort would be not flags and footprints or not *just* flags and footprints. See, for example, First DPT meeting, vol. I (24 June 1999), transcript, pp. 60, 79, 330, and especially 154–155.

"science driven, affordable, and engaging" into specific strategies and guidelines for action.

DPT's Charter

The team's basic guidance, its official charter, charged the group with developing a truly extended, long-term "vision" for the Agency. Although OMB provided funding for developing a 10-year plan, DPT's charter redefined "decadal planning" so as to allow the group to develop a plan that would guide NASA for the first 25 years of the new millennium. The charter's authors defined decadal planning "as the first ten-year definitive planning to reach a twenty-five year vision."[36] More importantly, the charter provided key concepts that the leaders of the Agency wanted the group to incorporate into their final plan. (See appendix D-1.)

The first characteristic the charter writers sought for the vision was for it to be "top-down," meaning that it should provide a coherent, focused plan for the NASA Administrator to implement, as opposed to a broad grab bag or laundry list of ideas that would allow priorities determined at the program, project, and Center levels to drive the Agency's mission in several unrelated directions. Such a "bottom-up" approach limited the possibility for coordination and priority setting.[37] The top-down approach offered the possibility that NASA could set a unified strategy that would guide the Agency for decades.

The deficiencies of past studies weighed on the minds of those charged with establishing the team. Garvin wanted to make sure that DPT did not follow the trajectory of other high-profile study groups whose legacy amounted to unrealized recommendations for moving the Agency in a bold new direction and a report relegated to the dustbin of history for only the most die-hard space enthusiasts to ponder.[38] As he told the assembled DPT members at the 24 June meeting, he spoke from experience, having worked with Dr. Sally Ride on her post-Challenger report intended to guide NASA's long-term strategy.[39] Released in August 1987, a year and a half after the Challenger accident, the Ride Report called for a variety of programs and plans, including establishing a human outpost on the Moon and then sending astronauts to Mars. Despite

---

36.   DPT charter, p. 1, NASA HRC.
37.   Thanks to Giulio Varsi for suggesting this line of thought.
38.   Garvin specifically noted that DPT should "NOT [be] another Synthesis Group Report or Mars Architecture Study" (emphasis in original). See Garvin, "First Meeting of Decadal Planning Team for Exploration," p. 5.
39.   First DPT meeting, vol. I (24 June 1999), transcript, p. 43. The formal name of the Ride Report was *NASA Leadership and America's Future in Space: A Report to the Administrator.*

the intense effort put into writing the report, it merely sat in libraries and on the bookshelves of longtime NASA employees like the reports of so many other blue-ribbon commissions and NASA working groups that proposed sending astronauts back to the Moon and on to Mars.[40]

How could DPT overcome the hurdles that prevented other proposals from inspiring the Agency's long-term exploration strategy? As noted earlier, to improve the team's chances for making a lasting impact, Goldin pushed for bringing on accomplished individuals who were young enough to implement and participate in the realization of the vision they would create.[41] Also, the group's charter called for the vision to be "forward looking and NOT tied to past concepts," meaning that although team members should take into account prior studies that held relevance for their work, they should create an original product that represented their most advanced thinking on the topic.[42]

The charter also contained specific ideas that would distinguish it from other exploration proposals. At the top of the list was the charter's requirement that DPT should produce a "science-driven, technology-enabled program development approach with technology roadmaps that enable capabilities at an affordable cost."[43] The team would later abbreviate this guidance with the catchy phrase, "science driven, technology enabled." What it meant exactly for the DPT vision to be "science driven" engendered a fair amount of discussion at the first meeting and at later points in time, in both formal and informal settings.

Some individuals at the June 1999 DPT meeting questioned whether science could be a sufficient "driver" for the vision. They greeted NASA claims of maintaining a scientific component on past human missions with skepticism. For example, Roger Crouch and Peter Curreri commented that they believed that NASA had overstated the scientific value of the International Space Station.[44]

---

40. More than 1,000 studies on how to send people to Mars were done in the second half of the 20th century. For an excellent overview of these studies, see Portree, *Humans to Mars*. A site called Key Documents in the History of Space Policy, found at *http://history.nasa.gov/spdocs. html* (accessed 19 April 2018), has links to a number of the more well-known studies of the 1980s and 1990s, including the Ride Report, the Paine Report, the Augustine Report, and the Stafford Report.

41. Guerra oral history, 26 October 2005, pp. 5–6.

42. DPT charter, p. 1.

43. Ibid.

44. First DPT meeting, vol. II (25 June 1999), transcript, pp. 53, 68. At that time, Crouch was working at Headquarters and Curreri had been detailed to Headquarters from MSFC's Space Science Lab. A Ph.D. physicist who had already flown twice aboard the Shuttle as a payload specialist, Crouch served from 1998 to 2000 as the Senior Scientist for the Office of Life and Microgravity Science and Applications. Perhaps ironically, Crouch then served as the Senior Scientist for the ISS from 2000 to 2004. See *http://www.jsc.nasa.gov/Bios/PS/crouch. html* (accessed 20 August 2008).

Informed observers have said the same about the Shuttle.[45] With the important exceptions of microgravity and biomedical research in orbit aboard the ISS and Shuttle, science has not been a central component of the human spaceflight program. In terms of advancing knowledge, the contributions of the ISS and Shuttle have been in the areas of engineering testing and development.[46] Thus, having a truly "science driven, technology enabled" vision would have represented a very significant historical shift.

Participants at the first meeting discussed the concept of science-driven and concluded that science was a necessary, but not sufficient, justification for exploration. While exploration may coincide with scientific inquiry, the two are not synonymous. As former NASA Chief Historian Steven J. Dick has pointed out, many of the polar explorers, as well as figures such as Columbus and Magellan, did not explore the regions for which they became famous to gain scientific understanding of our planet, although scientific knowledge was occasionally an ancillary benefit.[47] While not voiced explicitly at that time, one team member later opined that "exploration" is a "code word for humans beyond LEO."[48] Expressing skepticism about the potential impact of science on mission planning, Matt Golombek from JPL suggested that science goals are not necessarily plans, they are "just things you want to learn, things you want to know."[49] Mark Pine, a policy analyst and education outreach specialist from JPL, argued against "put[ting] everything under the umbrella of science."[50] Similarly, Scott Hubbard from Ames Research Center contended that the history of human exploration has been scattershot, so it would make sense to "pursue this grand

---

45. See, for example, Richard Monastersky, "Shuttle Science Called into Question Once More," *Chronicle of Higher Education* 49, no. 23 (14 February 2003): A17; Richard Muller, "Space Shuttle Science," *MIT Technology Review* (10 February 2003), at *https://www.technologyreview.com/s/401801/space-shuttle-science/* (accessed 8 June 2018), cited in Launius and McCurdy, p. 81; Robert L. Park, *Voodoo Science: The Road from Foolishness to Fraud* (Oxford and New York: Oxford University Press, 2000); Ted Katauskas, "Shuttle Science: Is It Paying Off?" *Research and Development* 32 (August 1990): 43–52; and Roger D. Launius, "Assessing the Legacy of the Space Shuttle," *Space Policy* 22 (2006): 226–234, especially 231.

46. The challenges of the Shuttle and ISS have shed light on basic problems that the space program might encounter elsewhere in the solar system, particularly with various spacecraft subsystems, such as electrical, computing, sanitation, hygiene, oxygen purification, and water processing. Technical failures have provided NASA program managers with numerous opportunities for learning how to identify, investigate, and ultimately solve life-threatening challenges in space. For various sources on the scientific value of the Shuttle, see the previous footnote.

47. On the differences between science and exploration, see Steve Dick's 11th "Why We Explore" essay, titled "Exploration, Discovery, and Science," at *http://www.nasa.gov/missions/solarsystem/Why_We_11.html*, 1 June 2005.

48. Dennis Bushnell, "Revolutionary Technology and Concepts" presentation, NASA HRC.

49. First DPT meeting, vol. I (24 June 1999), transcript, pp. 59–60.

50. First DPT meeting, vol. II (25 June 1999), transcript, p. 59.

adventure along multiple fronts."[51] Along these lines, Charles "Les" Johnson, a rocket propulsion specialist from Marshall Space Flight Center, noted that DPT should advocate "science-driven exploration, but not just exploration for the sake of science."[52] Jim Garvin concurred that science would not be the impetus for every individual element but that it would be a central justification for DPT.[53]

The DPT charter's only bolded sentence addressed this issue of the respective roles of science and exploration. The final sentence of the section enumerating the characteristics that the DPT vision should contain reads,

> This study should be viewed as the first small step toward a program designed to enable the inevitable and systematic migration of humans and robots into space beyond Earth orbit for the purposes of exploration, science, and commerce.

Thus, while the members of DPT understood that developing scientific rationales for exploration would be a central part of their task, the charter clearly identified the planning effort as just the first step in a long-term planning effort to project human and robotic capabilities beyond Earth's orbit, not just for scientific purposes, but for economic interests and to satisfy a basic human desire to explore as well.

Given the widespread perception that engineering priorities and goals have driven NASA's agenda historically, the prominent role assigned to science for the DPT vision is significant. Lisa Guerra recognized at the start that the team might have an uphill struggle to make science so preeminent at NASA,

---

51. First DPT meeting, vol. II (25 June 1999), transcript, p. 95; and presentation charts by Scott Hubbard, "NASA's Exploration Program: Issues for Discussion," from Early Brainstorming, APIO files, NASA HRC. Astronomer Neil de Grasse Tyson has argued that historically, there have been only three basic rationales for nation-states to explore: international prestige/ego of the national leader, economics, and national security. See his chapter "Expanding the Frontiers of Knowledge" in *Looking Backward, Looking Forward: Forty Years of U.S. Human Spaceflight Symposium*, ed. Stephen J. Garber (Washington, DC: NASA SP-2002-4107, 2002), pp. 127–136. Former NASA Chief Historian Roger Launius has added two other reasons for people to explore space: to fulfill human destiny and to address the eventual need for humans to live somewhere other than Earth. See, for example, his op-ed piece, "The Case for Humans in Space," *Space Times* (March–April 2002): 3.

52. See First DPT meeting, vol. II (25 June 1999), transcript, p. 52.

53. Ibid., p. 63. At another level, the famous mountaineer Sir Edmund Hillary noted that "[n]obody climbs mountains for scientific reasons. Science is used to raise money for the expeditions, but you really climb for the hell of it." This quotation can be found at *https://www. telegraph.co.uk/travel/destinations/asia/nepal/articles/quotes-sir-edmund-hillary-first-man-climb-everest/* (accessed 11 July 2018). Thanks to Giulio Varsi for pointing out this pithy quotation.

where the organizational culture has focused for decades on astronauts.[54] The challenge was twofold: to reorient NASA employees to focus on science and to explain to Congress and the public the value of doing so. Engineers and engineer-astronauts have dominated planning and decision-making at NASA since at least the early years of the Apollo program, when the urgency and difficulty of reaching the Moon led the Agency, in the minds of prominent scientists, to relegate science to a distinct second priority behind engineering.[55] The engineering and logistics components of the human spaceflight program have garnered the bulk of NASA's budget and attracted far greater public attention than other programs since the Agency's inception. Engineering prowess has remained critical to all space operations, even science-led programs, such as the Mars rovers and the Hubble Space Telescope. Astronauts, furthermore, have been the literal and figurative public face of the Agency. Over 70.7 percent of astronauts have experience and training in engineering disciplines, as opposed to scientific disciplines.[56] The differences between engineering and science, and between robots and humans in space, remained discussion points for the planning team throughout its existence.

Rather than reinvent the core "science drivers" for the vision, Jim Garvin suggested that the team draw upon existing lists of scientific priorities compiled by advisory bodies, such as the National Research Council (NRC).[57] The Decadal Planning Team did not need to venture beyond the NASA Office of Space Science for its scientific priorities. In the spring of 1998, Weiler's predecessor as Associate Administrator for Space Science, Wesley Huntress, asked his Division Directors to come up with fundamental scientific questions, so-called

---

54. First DPT meeting, vol. II (25 June 1999), transcript, p. 211.

55. Donald A. Beattie discusses some of the uphill battles that he and other scientists faced within NASA's engineering bureaucracy in trying to include scientific experiments on the Apollo missions in his book *Taking Science to the Moon: Lunar Experiments and the Apollo Program* (Baltimore: Johns Hopkins University Press, 2001).

56. Of the two main astronaut categories, pilots and mission specialists, over 90 percent of pilots are engineers, while approximately 53 percent of mission specialists are engineers. A number of mission specialists with a science background are medical doctors. See Duane Ross, "Astronaut Candidate Selection Process" presentation (2 November 2007), p. 14. This presentation uses the categories of engineering, physical science, biological science, and mathematics. Other similar data are in a chart that Bernadette Hajek, Technical Assistant to the Chief of the Astronaut Office at JSC, emailed to Steve Garber on 25 September 2008, showing a breakdown of current astronauts (in training or eligible for flight assignments). This chart breaks down the mission specialists' backgrounds into flight-test engineering, engineering and operations, physical sciences, life sciences, and education. These documents are deposited in the DPT/VSE, astronautsbackgrounds files in the NASA HRC.

57. The National Research Council's Space Studies Board, together with the NRC's Board on Physics and Astronomy, publishes formal decennial reviews and recommendations in astronomy and astrophysics that are well respected by scientists.

"Grand Challenges," in clear language that the public could easily understand.[58] The four space science Grand Challenges that Huntress and his division chiefs conceived called for NASA to read the history and destiny of the solar system, look for life elsewhere in the solar system, image and study extrasolar planets, and send a spacecraft to a nearby star. A fifth Grand Challenge was added for NASA's human spaceflight organization: to conduct a progressive and systematic program of human exploration beyond Earth orbit. Huntress gave a speech to the American Astronautical Society in November 1998 in which he outlined these Grand Challenges and called for a systematic, step-by-step approach to integrate human and robotic exploration of space beyond LEO.[59]

After Huntress left NASA in February 1998, his successor as Associate Administrator, Ed Weiler, broadened and refined the Grand Challenges with ideas from previous Office of Space Science strategic plans and advisory group recommendations. Weiler tried to simplify these challenges so that a layperson could understand them more readily. He posed them as four questions: how did the universe begin, how did we get here, where are we going, and are we alone?[60] At the first DPT meeting in June 1999, Harley Thronson urged team members to adopt these Grand Challenges as foundational concepts for their vision planning effort. (See Grand Challenges Document in appendix D-3.)

---

58. The practice of identifying "Grand Challenges" emerged in the late 1980s in the context of a broad national effort to maintain the U.S. lead in computer science and technology. With strong support from federal agencies, including the Defense Advanced Research Projects Agency (DARPA), NASA, the National Science Foundation (NSF), and the U.S. Air Force, for example, leaders in a range of scientific and engineering disciplines that depended on high-performance computers gathered in 1989 in Hawaii at "The Conference on Grand Challenges to Computational Science" to discuss current limitations in their fields and the computing breakthroughs they believed were needed for further advancement. The term "Grand Challenges" came into greater use in the early 1990s, as agencies across the federal government began to use the term to describe areas of research funded under the U.S. High Performance Computing and Communications Program. For background on early applications of the Grand Challenges concept, see Raj Reddy, "Foundations of Grand Challenges of Artificial Intelligence," *AI Magazine* (winter 1988): 9–21; Kenneth G. Wilson, "Grand Challenges to Computational Science," *Future Generation Computer Systems* 5 (1989): 171–189; and John H. Gibbons, Director of OSTP, "Information Infrastructure and HR1757, the 'High Performance Computing and High Speed Networking Applications Act of 1993,'" statement before the Committee on Science, Space, and Technology, House of Representatives, 103rd Cong., 1st sess., 27 April 1993.

59. See *The Planetary Report*, March/April 1999, pp. 4–6. See also a full copy of Huntress's speech titled "Grand Challenges for Space Exploration," NASA HRC. It is not immediately clear who added this fifth challenge or why. See appendix D.

60. Weiler interview, p. 17.

"Science Driven, Technology Enabled"

The DPT charter paired the requirement for creating a "science driven" vision with the expectation that the vision would recommend a "technology enabled program development approach."[61] What did the charter writers intend to convey with the term "technology enabled?" Technology has enabled every space mission since Sputnik, long before the 1990s, when it became fashionable to speak about "technology-enabled learning" and software producers first began using the term "technology-enabled selling" to distinguish Internet-based transactions (e-commerce) from face-to-face sales.[62] The opposite of a technology-enabled program is a program constrained by technology, not one lacking technology. The charter instructed the team members not to allow the current state of technology to limit their goals.

Presuming that new technologies would need to be developed in time, the charter requested that the vision include "technology roadmaps that enable capabilities at an affordable cost."[63] One reason for developing a roadmap, in the context of a major government technology initiative, is to provide associated researchers, institutions, and suppliers with a clear indication of the Agency's ambitions and a sense of the capabilities the Agency will likely need to achieve those ambitions. A roadmap would allow interested individuals and institutions an early start on developing new technologies and figuring out how existing technologies could achieve the capabilities needed to meet NASA's long-term exploration goals. As one member explained, the DPT vision should aim to bring about a "shift from technology derived from missions to missions enabled by technology."[64] A roadmap would allow NASA to avoid developing new technologies in the course of major missions, thereby shortening development times and reducing immediate project costs.

---

61. DPT charter, p. 1.

62. See, for example, "Calico Technology Announces Concinnity: First Sales Configuration and Quoting Solution To Use ActiveX," *Business Wire* (28 October 1996); and Jennifer Lewington, "Linking Schools with the Future Education," *Globe & Mail* (17 March 1994): A4.

63. DPT charter, p. 1. Although it was an extension of earlier practices and concepts that could be traced to the 1930s, technology roadmapping gained increasing acceptance as a planning technique in high-technology industries and government in the 1980s and 1990s. On the history of roadmapping, see Robert R. Schaller, "Technological Innovation in the Semiconductor Industry: A Case Study of the International Technology Roadmap for Semiconductors" (Ph.D. diss., George Mason University, April 2004).

64. John Mankins, "Technology for Human/Robotic Exploration and Development of Space Strategic Research and Technology Road Maps," December 2000, in DPT/VSE, DPT 2000–Gary Martin files, pp. 22, 24, NASA HRC.

Along these lines, Scott Hubbard suggested following the Defense Advanced Research Project Agency (DARPA) model, which involved providing seed funding in focused programmatic areas for technology development.[65] DARPA managers assumed that many of the projects they funded would not yield new technologies, but they anticipated that in most instances a small subset of projects would succeed spectacularly, yielding critical new technologies. The DARPA model was similar to the practice among venture capitalists of investing small sums in a large number of ventures to spread investment risks and increase the chances of success.[66] Even though NASA has supported research in a variety of science and engineering disciplines and has tried to foster commercial development of its technologies, the analogy is imperfect since NASA is an operating agency and DARPA is a research support arm of the Department of Defense.[67]

However, DPT members quickly realized that they should investigate certain key technologies, some of which had been problematic in the past. Three categories emerged clearly: launch vehicles, in-space propulsion, and nuclear power (for both electricity generation aboard spacecraft and spacecraft propulsion).[68]

NASA has long had an interest in technologies that improve upon existing in-space propulsion methods since traditional chemical rockets are expensive and complex and have reached their theoretical efficiency limit. So-called payload mass fraction is low, meaning that sending humans to Mars with today's technology would entail dozens of launches, most of which would simply carry rocket fuel. Two possible solutions to the efficiency problem are developing revolutionary new rockets that would be vastly more fuel-efficient or developing other techniques that use minimal or no propellant. Both approaches involve very slowly increasing a spacecraft's speed to produce low thrust for long-duration missions.[69]

---

65. See First DPT meeting, vol. II (25 June 1999), transcript, p. 206.

66. In mid-2003, NASA did establish a small venture-capital effort that received very little public attention. See Lisa Lacy, "Something Ventured: NASA Takes Off on Venture Funding," *Dow Jones Newsletters/Venture Wire* (23 December 2004). NASA's venture-capital effort was modeled on the Central Intelligence Agency's In-Q-Tel firm, which had been led by Dr. Michael Griffin, who later became the NASA Administrator.

67. For a good summary of DARPA's history, as well as how it operates, see Graham Warwick and Guy Norris's set of articles titled "DARPA at 50: Blue Sky Thinking," *Aviation Week and Space Technology* 18/25 (August 2008): 54–71.

68. DPT charter, p. 2. For a good summary distinguishing the two kinds of nuclear power in space, see W. Henry Lambright, "Federal Agency Strategies for Incorporating the Public in Decision-Making Processes: Case Studies for NASA," draft monograph, 18 April 2005, p. 46, available in the NASA HRC.

69. See "In-Space Propulsion Technology Program: Project Implementation Status," 26 July 2006, DPT/VSE, Les Johnson files, NASA HRC, p. 13.

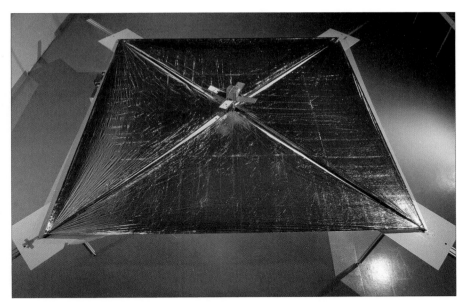

A solar sail, like this one, would use the force of the Sun's photons to propel a spacecraft. This sail is part of a NASA CubeSat called Nanosail-D. (NASA/MSFC/D. Higginbotham)

DPT members including Les Johnson looked at approaches such as electric propulsion, an example of radically more efficient systems that are still propellant-driven. Simply defined, this is the "acceleration of propellants by electrical heating, electric body [ion] forces, and/or magnetic body forces."[70]

Johnson knew that several kinds of "propellant-free" propulsion systems existed. One potentially promising technology was solar sails, which used the pressure of photons emitted from the Sun, rather than conventional propellants, to drive spacecraft. Another technology was momentum exchange tethers, which boosted spacecraft from Earth orbit to pre-escape trajectories. Aerocapture or aerobraking was a technique that harnessed a planet's atmosphere to slow an approaching spacecraft, allowing the spacecraft to enter a planetary orbit while expending virtually no propellant.[71]

In terms of nuclear systems, the team appeared to reach consensus early in its existence on the desirability of using nuclear power for long-range space

---

70. Karen Bishop-Behel and Les Johnson, "In-Space Propulsion for Science and Exploration" (paper presented at the National Space and Missile Materials Symposium, Seattle, WA, 24 June 2004), scan available in Les Johnson files, DPT, NASA HRC. The NASA Evolutionary Xenon Thruster (also known as NEXT), developed at NASA's Glenn Research Center, is an example of this ion propulsion technique.

71. "In-Space Propulsion Technology Program: Project Implementation Status," 26 July 2006, p. 10; and Bishop-Behel and Johnson, "In-Space Propulsion," pp. 17, 10 (both in NASA HRC).

activities. At the same time, the team recognized that fears about the risks of nuclear accidents would make selling nuclear-powered space exploration to political leaders and the public exceedingly difficult.

Garvin and the team knew that NASA would need nuclear Radioisotope Thermoelectric Generators (RTGs) to provide electricity for spacecraft to explore deep space. RTGs are nuclear batteries that harness the heat produced when highly radioactive plutonium decays to produce electricity in places where solar arrays would be impractical (they were used first on Earth and then successfully on various space missions). Despite their obscurity, RTGs have been in use since 1961. They are essential for long-duration missions where solar energy is not always available and where the requirements are high for reliability, power, longevity, and durability.[72]

A second potential use of nuclear energy in space is for in-space propulsion, using the fission of moderately radioactive uranium fuel. The advantage of nuclear fission propulsion is that it is approximately twice as efficient as that of conventional chemical rocket propulsion.[73] While nuclear fission propulsion in space has not been used by the United States, some ground tests have indicated that it may be feasible. Cutting lengthy trip times to Mars would lower astronaut exposure to radiation in space and would also reduce the chances of a serious system failure. Administrator Goldin was eager to cut trip times to reduce the effects of radiation exposure on the health of astronauts. Shorter trips would also reduce mission complexity and logistics.

Yet Goldin was keenly aware of the negative public connotations of the term "nuclear." He prodded the team to make the case for nuclear propulsion and nuclear-generated electricity as strong as possible so NASA would be prepared when political opposition arose.[74]

---

72. For Garvin's comments, see First DPT meeting, vol. I (24 June 1999), transcript, p. 39. For a good history of RTGs in space, see Roger D. Launius, "Powering Space Exploration: U.S. Space Nuclear Power, Public Perceptions, and Outer Planetary Probes," chap. 5 in *Historical Studies of the Societal Impact of Spaceflight*, ed. Steven J. Dick (Washington, DC: NASA SP-2015-4803, 2015), passim.

73. Lambright notes that NASA's Nuclear Engine for Rocket Vehicle Applications (NERVA) program, which extended from 1960 to 1972 and spent $1.5 billion in current-year dollars, was premised on the notion that after the success of Project Apollo, NASA would send astronauts to Mars. For more on NERVA, see James A. Dewar, *To the End of the Solar System: the Story of the Nuclear Rocket* (Lexington, KY: University Press of Kentucky, 2004).

74. See, for example, Harley Thronson's notebook from 20 January 2000, NASA HRC.

Human and Robotic Spaceflight

DPT also differed from previous study groups in that its charter explicitly called for the aggressive integration of human and robotic capabilities,[75] partly in an attempt to halt a contentious and long-running debate in the space policy community. Because of necessary safety and life-support functions, human spaceflight traditionally has been much more expensive than robotic exploration. Human spaceflight has been able to secure significantly higher funding than space science programs because of the inherently higher cost of human spaceflight, but also because of its popularity. The funding debate usually boiled down to whether these potentially competing budget priorities were in effect victims of a zero-sum game or whether increased budgets for human spaceflight created an environment favorable to higher budgeting for the space sciences and robotic exploration.[76] To resolve such tensions and because of the genuine belief that human and robotic capabilities would need to be integrated in the near future, the DPT charter endorsed a cooperative robotics-astronaut approach rather than calling for increased investment in both sectors (or favoring one over the other). A few DPT members, nonetheless, questioned whether humans truly were needed in space. While NASA has been flying astronauts for many years, the Agency has "no formal justification for their use other than construction/management of the International Space Station, except possibly for the purposes of public engagement in the space program."[77] Contrarily, Dennis Bushnell from Langley argued that without astronauts, NASA would be disbanded and its science functions given over to other agencies, although Bushnell believed that "there is no science justification for humans" to fly in space.[78]

Against the stereotype that the sciences were antagonistic toward human spaceflight, space scientists on the team forcefully promoted the notion that science would need humans in space. Thronson and Moe, for example, contended that astronauts "and their optimized robotic partners" would "become

---

75. DPT charter, p. 1. Humans have been controlling robotic spacecraft *from Earth* since spacecraft were first launched. The point is that this debate really concerns the relative scientific capabilities of robotics versus humans in situ (i.e., with astronauts in space). Thanks to Giulio Varsi for this insight.

76. Wesley T. Huntress, Jr., "Human Space Exploration Is About More Than Just Science," *Science* (8 August 2003): 771.

77. Charter for the Ad-Hoc Advisory Committee on Optimized Astronaut/Robotic Space Operations, September 2000, in Astronaut/Robot Review, Harley Thronson files, NASA HRC. In response to a challenge from then-Administrator Goldin in 2000, the Offices of Space Flight and Space Science formed a small team to work for several months to develop such a justification.

78. First DPT meeting, vol. I (24 June 1999), transcript, pp. 164–165.

indispensable beginning about 2010" because science goals "will require facilities and capabilities far too complex to construct and maintain without significant involvement of an optimized human/robotic 'workforce.'" They noted, however, that this contention was "by no means accepted throughout NASA, nor in the science and technology communities," and that the Agency had yet to devote enough resources to develop integrated astronaut-robotic capabilities.[79] Ultimately, the team opted for a compromise in the ongoing debate over human and robotic spaceflight, giving both a role in a new form of combined operations.

"Destination Independent" or Mars?

At the beginning of the first day, Garvin discussed the value of specific destinations in space as being tangible goals, and Mars as a readily appreciable goal for Congress and the public. Later that day, however, he suggested that the DPT charter was open-ended in terms of selecting destinations.[80] Before the two-day meeting was over, the team's focus had changed in the opposite direction and became *not* "destination driven." Lisa Guerra invoked the tag line from a Microsoft ad, "where do you want to go today?" to explain this philosophy. Thus in theory, NASA could develop generic technologies to enable astronauts to go to whatever destination seemed compelling at a particular time, whether it be the Moon, Mars, or Jupiter's moon Titan. The DPT charter did not explicitly use the phrase "destination driven"; rather, it called for a "vision of Space Exploration" for at least 25 years that would "open the human frontier beyond LEO by building infrastructure robotically at strategic outposts" such as libration points, planetary moons, and planets. The team did not propose a specific set of steppingstones such as successively sending people to Mars, Titan, and Pluto.[81]

The "destination independent" mantra held political hazards. First, destinations are priorities for those who oversee NASA in Congress. Without clear commitments to specific destinations, NASA likely would have a more difficult time convincing the taxpayers' representatives that the Agency planned to accomplish anything.[82] The team did intend to use various metrics such as

---

79. Harley Thronson and Rud Moe, "Optimizing Astronaut/Robotic Operations: Opportunities in the Time Period 2010–2020" in Astronaut/Robot Review, Harley Thronson files, pp. 2, 5, and passim, in NASA HRC.

80. First DPT meeting, vol. I (24 June 1999), transcript, p. 45. Garvin suggested that this was the direction he received from Rothenberg and Weiler. On p. 100, he suggests that the charter gives the team flexibility in terms of selecting destinations.

81. First DPT meeting, vol. II (25 June 1999), transcript, pp. 187–188. See p. 1 of charter.

82. First DPT meeting, vol. II (25 June 1999), transcript, p. 216.

safety, science, technology, schedule milestones, and cost to evaluate capabilities and progress in general.[83]

Also, the "destination independent" approach might draw criticism from spaceflight advocates who promoted focusing on specific destinations. Perhaps most prominently, Robert Zubrin advocated sending humans to Mars as quickly and inexpensively as possible.[84] Regardless of whether Zubrin and other Mars advocates held any influence, the task of explaining the goal of attaining overarching technological capabilities to "go anywhere, anytime" would be difficult.

While the DPT charter mentioned human exploration of planets, it did not specifically call for sending astronauts to Mars. The omission of explicit language calling for sending humans to Mars was a conscious decision on the part of the charter's drafters. The DPT charter writers framed the issue so that the team would not consider sending humans to Mars a foregone conclusion. The charter would allow the team to focus first on laying out the very highest-level scientific goals and then later consider the viability of conducting an ambitious robotic program followed by humans to Mars as a component of the larger scientific agenda.

As one scientist (not part of DPT) at another time insightfully asked: "If the answer is the Mars program…[with rovers and full-fledged human exploration and settlement, then] what is the question?"[85] In other words, what is so publicly compelling about Mars? Science may be a good reason to send humans to Mars, but to sustain a program over the long term, it must have a compelling rationale with significant political or economic consequences comparable to the Cold War competition with the Soviet Union that enabled the Apollo program to emerge as a top national priority. DPT member Les Johnson sensed that "there's an underlying current in this room of Mars as a preferred answer." He felt that the team needed to "start with a blank slate" and later posed the question "What can we do that has scientific merit, increases human reach into the solar system and is affordable?"[86] While sending people to Mars was (and still is) the implied/

---

83. The evaluation criteria section (pp. 2–3) of DPT's terms of reference was still rather general. (NASA HRC.)

84. See chapter 2 for more on Zubrin and the Mars Society.

85. Christopher McKay (presentation, "Workshop on Science and the Human Exploration of Mars" proceedings, NASA Goddard Spaceflight Center, sponsored by Codes S and M, organized and managed by the Lunar and Planetary Institute, 11–12 January 2001), pp. 45–46, in Miscellaneous—Harley Thronson files, NASA HRC. McKay is a planetary scientist at NASA Ames Research Center (see *https://spacescience.arc.nasa.gov/basalt/people/christopher-mckay/*, accessed 12 July 2018).

86. First DPT meeting, vol. II (25 June 1999), transcript, p. 84; "Personal View: Exploration Strategies" (undated one-page memo, "typed and attributed to Les Johnson") from Early Brainstorming folder, APIO files, NASA HRC.

de facto goal for many NASA employees (and many Americans) for various cultural, scientific, technological, and other reasons, the team members believed they would need to defend publicly any decision to make Mars an explicit goal.

Mars holds a special place in American culture. As the target of a rich history of science fiction and fantasy, Mars resonates with the general public as a destination for human exploration to a far greater extent than other planets.[87] While much further away than humans have traveled previously, Mars is also much closer to Earth, and it is easier to reach than most other places in our solar system. Scientists who study and advocate for Mars exploration, furthermore, are well entrenched and well respected in the space community. They believe that exploration of the Red Planet would be rewarding scientifically and that the technical and financial obstacles to Mars exploration could be overcome someday. Thus vocal advocates for Martian exploration exist not only in U.S. popular culture, but within the robotic (science) and human spaceflight communities.

By a process of elimination, some in the space community contended that Mars was the only feasible destination for astronauts to travel within the next 50 years or so. Mercury and Venus are too hot, Jupiter and the outer planets are too far and cold, we have already visited the Moon, asteroids would be challenging and probably not as rewarding scientifically as Mars, and Lagrangian points are too esoteric for the public to appreciate.[88]

Funding

In terms of funding, the DPT charter called for "buying by the yard," a phrase indicating that the affordability of the program trumped other priorities. The charter, in other words, recommended that budgets determine the pace of investment in new technologies, generic technological capabilities, infrastructure, and scientific research, rather than allowing unrealistic timetables or performance expectations to drive budget requests.[89] The charter likely took a cautious position on financing in the hopes of avoiding the disdain and disbelief that greeted the price tag of approximately half a trillion dollars that accompanied the Space Exploration Initiative in 1989. The Augustine Report of 1990 had advocated a go-as-you-pay approach in reaction to SEI's projected huge costs, although,

---

87. On American cultural fascination with Mars, see the Howard E. McCurdy, "Mysteries of Life," in *Space and the American Imagination* (Washington, DC: Smithsonian Institution Press, 1997) pp. 109–138.

88. Thanks to Giulio Varsi for making this general point. In a noted 2010 speech at Kennedy Space Center, former President Barack Obama called for sending astronauts to an asteroid. (See *http://history.nasa.gov/Obama%20speech%204-15-10.pdf*, accessed 23 April 2018).

89. DPT charter, p. 2.

as noted in the previous chapter, the report also advocated increasing NASA's budget by 10 percent per year over a decade, after adjusting for inflation.[90] The go-as-you-pay model for R&D came with hazards for NASA. Many advocates of space exploration, as well as other science and technology efforts, have long lamented the inability of Congress to provide stable funding for long-term projects, even when NASA's overall budget is relatively stable. It can be difficult for the European Space Agency to cooperate with NASA on robotic space missions, for example, when the former's budget cycle is five years and the latter's is annual at best (given congressional continuing resolutions and even sequestration).[91]

"Buying by the yard" was related to the concept of technological "steppingstones," a common theme of many planning studies of the last half century.[92] To give DPT members perspective on what had come before them, Lisa Guerra summarized earlier studies that considered the future of NASA and the space program. She explained that studies such as *America at the Threshold* (the "Stafford Report") of 1991 provided more options than the single path outlined in the 90-Day Study that immediately followed SEI's announcement in 1989. The Stafford Report also called for sending humans to the Moon and Mars with a steppingstones approach and an emphasis on science.[93] At the June 1999 kickoff meeting, Harley Thronson relayed Ed Weiler's comparison of the interstate highway system, with its infrastructure of connecting bridges, truck stops, restaurants, and so forth, to robotic and human exploration of space, which likely would need infrastructure at Lagrangian points, the Moon, Mars, and other locations.[94]

---

90. See, for example, a summary of SEI by former NASA Chief Historian Steve Dick at *http://history.nasa.gov/seisummary.htm* (accessed 23 April 2018). For more discussion of the Augustine Commission and specifically Augustine's personal philosophy on go-as-you-pay, please see chapter 2. The Augustine Report is available at *http://history.nasa.gov/augustine/racfup1.htm*.

91. The various permutations of the ExoMars project represent one example of this difficulty caused by different budgetary horizons. See, for example, *https://mars.nasa.gov/programmissions/missions/present/esa-2016-exomars-trace-gas-orbiter/* and *http://www.esa.int/Our_Activities/Space_Science/ExoMars* (both accessed 15 June 2018).

92. Webster's Third International Dictionary, Unabridged (Springfield, MA: Merriam-Webster, Inc., 1986) defines this term as "a means of progress or advancement (the law was a steppingstone to a career in politics)." In the context of DPT, it meant a logical sequence of progressively more challenging tasks to build technological or scientific capabilities.

93. First DPT meeting, vol. I (24 June 1999), transcript, pp. 389–393. Please see chapter 2 for a brief discussion of the 90-Day Study.

94. First DPT meeting, vol. I (24 June 1999), transcript, p. 122. Lagrangian or libration points are positions in space where a spacecraft may be placed and remain almost stationary with respect to Earth and the Moon. The reason such points exist is because three effects cancel each other out: two gravity fields and centrifugal force. Five such points exist. See, for example, *http://www.nas.nasa.gov/About/Education/SpaceSettlement/CoEvolutionBook/3D.html* (accessed

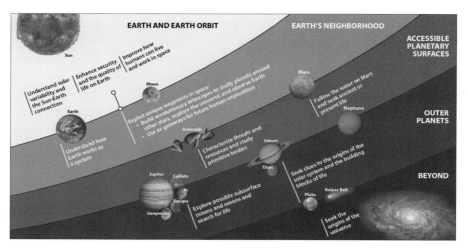

DPT illustration of the steppingstones concept. (NASA, from Gary Martin-electronicfiles\NEXT Doc Portal CD1\NEXT_2002\NEXTH.PDF 2/12/02, p. 22)

Where would the money come from to implement DPT's ambitious plans? The team initially discussed using anticipated future savings (a so-called "wedge") from NASA's human spaceflight program to fund the broad-scaled exploration investments DPT advocated.[95] The budget for space station development and operations amounted to approximately $2 billion in the early years of the program, up to the start of DPT.[96] Conceivably, at the time, between the expected savings from the Shuttle and Space Station Programs, a total of $2 billion could have been redirected to new space exploration efforts. Some experienced NASA insiders believed that costs for the station would drop to

---

19 September 2005). Some astronomers are planning large telescopes at these Sun-Earth Lagrangian points because their position would be stable and telescopes in space do not suffer from the effects of Earth's atmosphere, as ground-based observatories do. A spacecraft requires little energy to shift between libration points, and "access to all location on [the] moon and Mars is equivalent." (See p. 23 of "Exploration Blueprint Input: Integrated Space Plan," 21 November 2002. in Space Architect files, NASA HRC). For example, the James Webb Space Telescope will be launched to a Lagrangian point, and the Solar and Heliospheric Observatory (SOHO) and Wilkinson Microwave Anisotropy Probe (WMAP) spacecraft have already operated there successfully.

95.   For discussion of this "wedge," see Lisa Guerra's comments on p. 61 of the first DPT meeting, vol. I (24 June 1999), transcript.

96.   See tables 7-13A, 7-13B, 7-14A, and 7-14B of Judith A. Rumerman, *NASA Historical Data Book, Vol. VIII: NASA Space Applications, Aeronautics and Space Research and Technology, Tracking and Data Acquisition/Support Operations, Commercial Programs, and Resources, 1989–1998* (Washington, DC: NASA SP-2012-4012, 2012) (available at *http://history.nasa.gov/databooksvol8/NASA_Historical_Data_Book_8_Tables.pdf*); and p. 6 of Guerra oral history, 2 November 2005, NASA HRC.

$0.5 billion per year once the program got under way in earnest, with regular U.S. flights to the Station. This notion was abandoned quickly when, in 2000, little more than a year after the start of DPT, major cost overruns on the ISS dashed all hopes for savings from human space operations.[97]

DPT members at the June 1999 meeting expressed skepticism that the ambitious plans they were beginning to develop could be implemented within current budget projections, even with the $2 billion expected savings from the Shuttle and Station Programs. Eventually, the program would encounter a "budget lump," using the analogy of a snake swallowing a large animal, that would require additional funding. Conducting extensive robotic exploration of Mars and employing Huntress's and Weiler's Grand Challenges strategies of conducting development at a measured pace could minimize initial costs, but sending humans to Mars would necessitate a significantly expanded budget. The Augustine Report, as Lisa Guerra noted at the first DPT meeting, suggested that NASA needed a 10 percent budget increase in real terms to accomplish any major new program that included sending humans to Mars.[98]

Don Pettit from JSC talked about the "ghosts of studies past" that did not lack vision, science goals, or technology paths, but all fell prey to the problem of how to line up political support for a program that would entail major budget increases for NASA. DPT thinking about sending humans to Mars was based in part on the seminal Mars Reference Mission study of 1997.[99] In turn, this study adopted the "live off the land" philosophy of in situ resource utilization that Robert Zubrin had espoused to make human travel to Mars less costly. The Mars Reference Mission authors wrote that they wanted to "challenge the notion that the human exploration of Mars is a 30-year program that will cost hundreds of billions of dollars."[100] Pettit questioned whether he and his DPT colleagues could expect much political support. In agreement with the go-as-you-pay philosophy, Pettit recommended focusing on an open-ended

---

97. See pp. 6–8 of Guerra oral history, 2 November 2005.

98. First DPT meeting, vol. I (24 June 1999), transcript, pp. 313, 316, 321, 387.

99. Stephen J. Hoffman and David I. Kaplan, eds., *Human Exploration of Mars: The Reference Mission of the NASA Mars Exploration Study Team* (Houston: Johnson Space Center, NASA SP-6107, July 1997).

100. See First DPT meeting, vol. II (25 June 1999), transcript, p. 121, for Pettit's thoughts on political support for a multibillion-dollar program. See also Hoffman and Kaplan, *Human Exploration of Mars*. The quotation is from page 1-6. On page 1-3, this publication explicitly notes its foundations in Zubrin's work and in the 1991 Synthesis Group Report (at *http://history.nasa.gov/staffordrep/main_toc.pdf*, accessed 23 April 2018). Zubrin had given a presentation on his "Mars Direct" strategy in January 1991 at a meeting of the American Institute of Astronautics and Aeronautics. Zubrin also published his thoughts in *The Case for Mars* (Simon and Schuster, 1996).

program and consciously avoiding providing a total cost figure that would draw the attention of Congress.[101]

To persuade Congress and the American people to support a new exploration program, the team believed, NASA would need to explain its goals using concise and simple language. DPT members focused on engaging the public with a concerted communications strategy that conveyed concepts the average layperson could grasp with little effort, as had occurred with the Apollo program.[102] Huntress's and Weiler's iterations of the Grand Challenges were other attempts to use plain and compelling language to describe NASA's science mission.

In terms of U.S. constituencies, DPT members thought that NASA might seek to entice private corporations to get involved in the program at an early stage to share costs. The charter said little about industry other than distinguishing the responsibilities of government and private business in space, claiming that "industry has commercialized LEO [low-Earth orbit] and GEO [geosynchronous orbit]—NASA's job is to expand the frontier beyond Earth orbit."[103] With this statement, the charter writers meant to convey the conventional wisdom that the federal government's key role in science and technology is to push the frontiers of research and technology, particularly in areas that the private sector avoided because of a lack of clear financial rewards. As in the case of satellite communications, NASA's job was to assist in establishing a new business activity and then move out of the way to let industry handle commercialization, extract profits, and develop further capabilities. Paul Westmeyer recommended moving quickly to get industry involved at an early stage of the program, rather than handing off to industry at a later date or wrestling over roles and responsibilities.[104]

Embargoed in the Near Term

The team was determined not to let its effort result in the production of yet another government committee report that provoked few controversies and sat on the bookshelves of NASA leaders without leading to any serious effort to implement the recommendations contained within. They wanted to make a truly lasting impact and help shift the human spaceflight program toward what many in the public would consider more meaningful and ambitious activities. They wanted to avoid building a program focused primarily on accomplishing

---

101. See First DPT meeting, vol. II (25 June 1999), transcript, p. 124.
102. Ibid., pp. 123, 150–151.
103. DPT charter, p. 2.
104. First DPT meeting, vol. I (24 June 1999), transcript, pp. 138–139, 144.

symbolic goals for political purposes. Throughout the first day of their initial meeting, they repeatedly expressed a shared belief that science combined with a continued, sustained human presence in space, particularly at venues such as Mars or one of the Lagrangian points, could be the guiding principles that would make the results of the DPT planning effort truly influential.[105]

The DPT charter established the parameters for the team's work and set the group apart from previous planning efforts. The charter directed the team to create a plan to answer basic scientific questions using capabilities enabled by an incremental technology development program and the aggressive integration of humans and robotics in space. While other studies, such as the Paine and Augustine Commission Reports, called for strong support of space science, DPT was unique in that its charter called for science to serve as the fundamental rationale for human exploration. Conceived with the perceived mistakes of the past in mind, DPT's charter also called for the group to take a tack different from that of other blue-ribbon commissions and study groups, particularly those that called for bold programs requiring large investments. The DPT charter created an expectation that the team's plan should be constructed with distant horizons in mind. Rather than develop a plan for NASA to sell to Congress and the American public as a single, coherent package with an exact price tag, the charter called for DPT to create a plan that focused on incremental investments in technology and allowed NASA to pursue its goals regardless of short and near-term vacillations in funding.

DPT also differed from past studies in how it operated. Administrator Goldin asked the team to embargo its deliberations and work products. In practical terms, this meant that Goldin expected team members not to discuss their deliberations or share their work products with anyone not involved directly in DPT activities. Major blue-ribbon commission studies typically were headed and staffed by non-NASA personnel, but DPT was done "in-house" with representation from each Center and NASA Headquarters.

The DPT planning effort and its follow-on activities received modest financial support. The team first assembled without direct funds. Funding began with the start of fiscal year 2000, in October 1999. The $5 million annual amount allotted for 2000 dropped to approximately $4 million annually for fiscal years 2001–2004. In January 2000, Weiler and Rothenberg urged the team to spend

---

105. Jim Garvin told the team that it was to focus on "brainstorming" and did not necessarily need to produce a standard report. See First DPT meeting, vol. I (24 June 1999), transcript, pp. 18, 46. For the brief discussion about DPT's broad mandate, see p. 53 of the same document. For the discussion of avoiding "flags and footprints," see pp. 54, 79, 154–156, 330 of the same document.

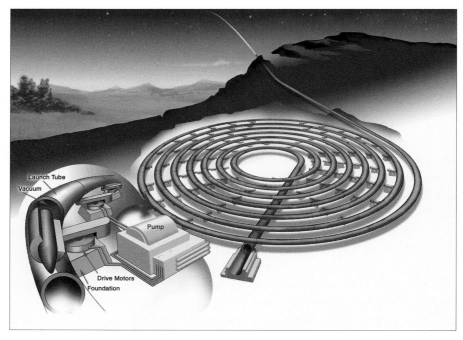

A slingatron from a 22 November 2000 DPT presentation. Slingatrons and blast wave accelerators were potential alternatives to chemical rockets as launch technologies. (NASA from Lisa Guerra-electronic files\Key DPT & Next Briefings\\Exploration Planning.pdf 4/18/05, p. 6)

their limited funds, but to do so carefully so as not to call undue attention to DPT and risk having funding pulled away to solve other, bigger NASA financial problems.[106] The planning effort did not require the purchase of hardware or funds for maintaining a large workforce. The group spent its funding mostly on feasibility studies for technologies such as slingatrons, blast wave accelerators, and carbon nanotubes, as well as on travel, meetings, and reimbursement of some Center personnel costs.[107]

Although the team's work was tightly embargoed initially, Goldin spoke freely about the human exploration of Mars in public throughout most of his tenure. During his first speech to all Headquarters employees as Administrator in June 1992, Goldin picked up on President George H. W. Bush's Space

---

106. For the specific annual budget figures for DPT, see the "Next Decade Planning Funding History" document in budget materials obtained from Harley Thronson, 15 September 2005. For discussions of assumption of a flat DPT budget, see, for example, First DPT meeting, vol. II (25 June 1999), transcript, pp. 13, 69, 100. For Weiler and Rothenberg's comments on DPT's budget, see Decadal Planning Team–2 Retreat Report, 20–21 January 2000, St. Michaels, MD, NASA HRC.

107. See DPT Budget spreadsheet in Miscellaneous Cost, Lisa Guerra files, NASA HRC.

Exploration Initiative and called for sending astronauts back to the Moon "to <u>stay</u>. [emphasis in original] And before the 50th anniversary of Apollo 11 [2019], we'll plant the American flag on Mars." In a July 1995 speech, Goldin called for a "sustained presence on Mars, not a one-shot, spectacular mission. Emigration, not invasion." In May 1997, Goldin told a National Space Club audience that "$2 billion a year gets us to Mars with sustained presence before the end of the first decade of the next century." At a 1998 gala event commemorating NASA's 40th anniversary, Goldin employed a frequent trope of his, talking about having a female astronaut step onto Mars in what he hoped would be much less than 40 years because "[i]n 40 years…we will have a colony on the Red Planet." In a speech about Mars to the National Geographic Society in 2001, Goldin spoke about going to the Hayden Planetarium in New York City as a young boy and how it inspired him to become an engineer and then accept the top NASA job: "My dream was—and still is—to go to Mars. And to be perfectly honest, I didn't think we were moving anywhere near fast enough [before he became Administrator] to get there."[108] Despite his regular comments in support of human exploration of Mars, Goldin did not reveal that he was doing anything to make it happen. He hoped to "sneak up on Mars" by keeping the details of DPT's deliberations quiet until the most opportune time from a political standpoint.

Although Goldin's direct involvement in DPT lasted only about 18 months, from the initial kickoff meeting at his house in April 1999 through October 2000, he deserves credit for helping to start this long-term planning process. He bridged the human and robotic spaceflight communities within NASA by appointing two friendly colleagues, Rothenberg and Weiler, as stakeholders. By initiating and supporting a relatively novel attempt to develop human spaceflight plans based on scientific goals, Goldin cut a path for those in the space community who worried that the human spaceflight program had no easily identifiable purpose. He raised the possibility that NASA could once again hold the public's attention with a spaceflight program that would inspire future generations, lead advances in a range of scientific and technical fields, and gain sustained political support. In short, he provided an opportunity for talented mid-career scientists and engineers within NASA to think in creative ways about the future of human spaceflight in preparation for opportunities that might lie ahead, in the next presidential administration or beyond. The first phase of DPT involving brainstorming, based in part on studies of exotic technologies connected to

---

108. These speech quotations are pulled from document record numbers 32052, 31449, 37612, 31906, and 33539, respectively, in the NASA HRC database.

scientific drivers, to set a strategic direction and find a compelling rationale for human spaceflight, continued through the end of 1999.

Phase Two Begins

The second phase of DPT formally began in January 2000. This phase was intended to focus the earlier brainstorming efforts on specific design studies to assess different technical capabilities and to flesh out the different architectures that would be needed.[109] The team started to hold meetings more frequently and increase the intensity of its effort.[110] During 2000, the team started to grapple in earnest with key technical issues involving human-robotic exploration of space beyond LEO. DPT members also started to focus on how best to portray their work when the time came to make it public.

In a private meeting with some of the DPT members on 19 July 2000, Administrator Goldin critiqued the team's effort and gave recommendations for improving its ideas and plans. He pushed the team to make its strategy more explicit and to create a continuum of logical destinations including the Moon, libration points, and Mars. He was concerned about "getting stuck on the Moon," and urged the team to "[g]et me to Mars!"[111] He emphasized that NASA would need significantly new infrastructure in many areas, such as heavy-lift launch vehicle technology, space communications, and space reconnaissance, to get humans to Mars repeatedly.[112] On the biomedical side, he decried the lack of progress on radiation effects research and pushed for more creative studies in this area. While he believed that nuclear fission propulsion might be necessary from a technical standpoint to reach Mars, he worried that

---

109. NASA Exploration Team, "Setting a Course for Space Exploration in the 21st Century: June 1999 Through December 2000," report, 25 February 2002, p. 8, in Lisa Guerra files, folder 7, NASA HRC.

110. In 2000, the DPT team met eight times. See "DPT Meetings in 2000" document in Nguyen, DPT Meetings files, NASA HRC.

111. Harley Thronson's notebook, entry for 19 July 2000, GSFC 2002 folder, NASA HRC. This meeting also had some colorful discussion of the "sneaking up on Mars" idea.

112. Heavy-lift launch vehicles for cargo are known colloquially as "big, dumb boosters." This concept stems from the notion that launch costs for robotic cargo could be minimized by utilizing relatively simple, expendable rockets. In the early 1960s, rocket pioneer and iconoclast Robert Truax proposed a big, dumb booster that he dubbed the Sea Dragon. See *http://ntrs.nasa.gov/archive/nasa/casi.ntrs.nasa.gov/19880069339_1988069339.pdf* (accessed 13 February 2009). See also, for example, Arthur Schnitt and F. Kniss, "Proposed Minimum Cost Space Launch Vehicle System," Aerospace Corporation, TOR-0158(3415)-1, 18 July 1966; and "Big Dumb Boosters: A Low-Cost Space Transportation Option?" (Washington, DC: International Security and Commerce Program, Office of Technology Assessment, U.S. Congress, 1989), *http://www.princeton.edu/~ota/disk1/1989/8904/890403.pdf* (accessed 13 February 2009).

the public would never support it. Finally, Goldin expressed concern that the budget for such grand plans could be unattainable and suggested that the group present its ideas and budget numbers to a "murder board" (a panel of reviewers who would ask difficult or critical questions) that included Sam Venneri, NASA's Chief Technologist, and Mal Peterson, the Comptroller. From his position as the guardian of NASA's purse strings, Peterson asked questions that would prepare the team to face even tougher challenges in the future from the administration, Congress, journalists, and the public.[113]

Wye River: A Turning Point

On Thursday and Friday, 12–13 October 2000, a year and a half after its formation, the DPT team held a retreat in Wye River, Maryland, that turned out to be a turning point, but not in the direction it had hoped. Goldin, as well as Rothenberg and Weiler and Comptroller Mal Peterson, attended the meeting, but the Administrator was distracted by other Agency issues that had little direct relationship to DPT's work. Several team members recalled the meeting as a low point, with one person calling it "abysmal."[114] Goldin went on tirades about a variety of unrelated subjects, including NASA's parochial managers who were unduly loyal to their Centers or programs, lack of technological progress, and ISS cost overruns. He complained about delays for the launch of the first ISS crew, which entered space later that month as Expedition 1, and expressed dissatisfaction with the state of space medicine research and the ability of NASA to ensure the health of the ISS crew. Goldin also pushed DPT to come up with more tangible objectives to get away from an "entitlement mindset" and to explain why astronauts were so critical to space exploration. He fumed about hearing tired rhetoric rather than seeing real progress on supposedly cutting-edge technologies, such as space nuclear power and fission propulsion, as well as alternative launch vehicles. He also cautioned the team, however, that NASA could not recommend any new nuclear technologies yet because of the public's negative attitude toward projects involving nuclear materials. While Goldin indicated that he liked aspects of some of the presentations, such as one on a plasma field solar sail, and he even conceded that he was heavy-handed, Goldin made clear that he thought the team and NASA had a

---

113. Harley Thronson's notebook, entry for 19 July 2000.
114. Guerra interview, 26 October 2005, pp. 61-64; Harley Thronson, interview by Glen Asner and Stephen Garber, NASA Headquarters, Washington, DC, 3 March 2006, pp. 26–30; Martin, interview by Asner and Garber, NASA Headquarters, Washington, DC, 16 February 2006, pp. 51–53 (all in NASA HRC). The "abysmal" quotation is from the Guerra interview, p. 61.

lot of work ahead, particularly in new launch technologies, space medicine, and human-robotic synergy. Rothenberg and Weiler participated in the discussion, but Goldin clearly dominated the meeting.[115]

Beyond all these potentially troubling issues, another cloud was hanging over Goldin's head. At the time, Goldin was one of only a few NASA people who knew that the ISS was experiencing an estimated $4 billion dollar cost overrun.[116] While not related to DPT, this information weighed heavily on the Administrator, who fought hard for the ISS Program from its inception. Although he claimed after the fact that cost overruns were routine and had little bearing on his disposition or decisions at the meeting, DPT members felt otherwise. They thought he was distracted, at best, and maybe even furious about the problems weighing on his mind.[117] The DPT team was dispirited after this meeting, in marked contrast to their enthusiastic start 18 months earlier.

It also seemed clear that Goldin had lost interest in DPT by the Wye meeting. Early on, he enjoyed meeting with DPT and even joked that he wished he could be a member of the team.[118] But after the Wye River meeting, he was preoccupied with more pressing problems with ongoing programs and no longer had time for DPT-style brainstorming. Some participants came to believe that Goldin wanted to disband DPT after the Wye meeting. For this reason, the DPT team took an even lower profile within NASA. Team members did not recall meeting with Goldin again about DPT after October 2000. Goldin retrospectively did not recall the Wye retreat nearly as negatively as the other participants and did not remember what happened to DPT after this time period.[119]

One positive result of this meeting was that the in-space propulsion effort of Marshall Space Flight Center's Les Johnson was given a higher profile and more funding. Johnson gave a presentation at the meeting in which he described advanced in-space propulsion technologies as potentially revolutionary,

---

115. "RWE" (Robert Easter), JPL, "Notes from DPT Wye River Retreat 10/12 & 10/13, 2000," 16 October 2000, in DPT 2000 folder in Gary Martin files, NASA HRC. These "RWE" notes provide a colorful and candid account of the meeting.

116. This $4 billion overrun, which was projected to be spread over five years, seemed to appear unexpectedly on Capitol Hill and in the public eye. Congress had enacted a cost cap on the ISS in 2000, and then the House Science Committee held a hearing on this issue on 4 April 2001. Goldin and Marcia Smith from the Congressional Research Service testified, as did a member of the ISS Cost Assessment and Validation Task Force (see *http://history.nasa.gov/32999.pdf* [accessed 25 April 2018] for a copy of this group's 1998 "Chabrow Report"). The testimony of these people, as well as some related media reports, are in file 17089, NASA HRC.

117. Goldin interview, pp. 27–28.

118. Guerra interview, 26 October 2005, p. 61.

119. Guerra interview, 26 October 2005, pp. 61–64; Thronson interview, pp. 26–30; Martin interview, 16 February 2006, pp. 51–53; and Goldin interview, pp. 29–31.

particularly for robotic spacecraft. Johnson led an in-space transportation program at the Center level for NASA's Office of Aerospace (or Aero-Space at some times) Technology and was selected for DPT because of his expertise in this area. In his role as a DPT participant, Johnson advocated strongly for the goals of the project, formally known as the In-Space Transportation Investment Area, which was funded at $3 million for FY 1999–2001.[120]

Shortly after the Wye River retreat, the Office of Space Science's Solar System Exploration Division assumed responsibility from the Office of Aerospace Technology for this effort, starting in FY 2002 at a budget of $19.5 million. After Sean O'Keefe succeeded Dan Goldin as the NASA Administrator in December 2001, all nuclear technology efforts, including those for in-space propulsion, were transferred to the newly formalized Project Prometheus (focused on nuclear systems to generate electricity for on-board power and propulsion) in FY 2003.[121] Although the budget for the project would continue to grow, reaching $26.5 million in FY 2006, budget cuts in later years (for FY 2008–2012) eliminated nearly all funding for testing and technology development.[122] Support for the project following the Wye River meeting was short-lived, but it allowed Johnson and his team to demonstrate tangible results with highly experimental propulsion technologies, including solar sails, aerocapture shells, and ion engines.

Following Wye River, DPT continued to work without direct contact with the NASA Administrator, a situation that some team members referred to in

---

120. "In-Space Propulsion Technology Program: Project Implementation Status," 26 July 2006, p. 6.

121. Johnson, telephone conversation with Steve Garber, 1 April 2009, notes in Emails to Authors, DPT Collection, NASA HRC; "In-Space Propulsion Technology Program: Project Implementation Status," 26 July 2006, p. 6; Bishop-Behel and Johnson, "In-Space Propulsion for Science and Exploration," p. 4; Les Johnson, "In-Space Transportation Technologies Overview," 13 March 2001, ("5792_InSpace_Hartman_Mar15.pdf" file in "LesJohnson-inspacepropulsion" electronic files), NASA HRC; *Project Prometheus: Final Report* (Pasadena, CA: NASA Jet Propulsion Laboratory, 1 October 2005), available at *https://trs.jpl.nasa.gov/bitstream/handle/2014/38185/05-3441.pdf?sequence=1&isAllowed=y* (accessed 12 June 2018). For more on Sean O'Keefe, please see his entry in the biographical appendix.

122. "In-Space Propulsion Technology Program: Project Implementation Status," pp. 32, 38, 39; Rae Ann Meyer, Deputy Manager, In Space Propulsion Technology Projects Office, Marshall Space Flight Center, "In Space Propulsion" presentation, 13 April 2004 ("ISPOverview041304FluidsWorkshop.pptx"), NASA HRC, pp. 3, 5; Bishop-Behel and Johnson, "In-Space Propulsion for Science and Exploration," p. 3; Les Johnson, "In-Space Transportation Technologies Overview," p. 5; Johnson, telephone conversation, 1 April 2009. Deep Space 1, a technology demonstrator spacecraft that launched in 1998, had successfully demonstrated the first use of ion propulsion, in which xenon gas is ionized by electron bombardment. See *https://www.jpl.nasa.gov/missions/deep-space-1-ds1/* (accessed 12 June 2018). For a more detailed DPT presentation on in-space technologies, see Les Johnson, Rae Ann Meyer, and Randy Baggett, "In-Space Transportation Technologies Overview," 18 April 2001, NASA HRC.

jest as "sneaking up on Dan Goldin." They did not intend to circumvent Goldin directly, and he never ordered that their work cease. Weiler and Rothenberg encouraged the team to continue, taking a slightly different tack.[123]

DPT and the 2000 Presidential Election

The strange paralysis that gripped the American public and political system as the nation waited for the 2000 presidential election results had little impact on NASA or the Decadal Planning Team. As discussed earlier, Goldin made no pretensions about having a political crystal ball. He created DPT to provide a new presidential administration with a range of options for space, no matter who the new president would be. DPT simply continued its work while the state of Florida and the Supreme Court sorted out the election results.

On 19 December 2000, about a week after Al Gore conceded the election to George W. Bush, DPT gave a presentation to members of the Office of Management and Budget at NASA Headquarters. Steve Isakowitz, the lead budget examiner for NASA, attended the meeting. Jim Garvin gave a memorable pitch. He started out by asking to speak uninterrupted for 15 minutes before providing details and allowing other team members to speak. The meeting, according to participants, went far better than the October Wye River retreat with Dan Goldin.[124]

From his positions at OMB and NASA, Steve Isakowitz was instrumental in initiating the decadal planning effort and in developing the VSE. (NASA)

Garvin began the presentation with a discussion of antecedents, which Thronson and other teammates had urged on the team to encourage a historical mindset. Garvin explained how President Thomas Jefferson had supported the Lewis and Clark expedition at the beginning of the 19th century and President Roosevelt had encouraged the nation to explore nascent aircraft flight at the beginning of the 20th century. In addition to discussing the group's plans for a destination-independent program that relied on steppingstone technologies, Garvin posed three of Weiler's "exploration Grand

123. Thronson interview, pp. 30–31.
124. Garvin interview, pp. 52–53.

Challenges" (how did we get here, where are we going, and are we alone?) (see appendix D). To underscore the importance of human-robotics synergy in space, DPT used the most well-known example: in 1993, astronauts serviced the Hubble Space Telescope and successfully installed a corrective optics package, enabling the HST to be a world-class scientific instrument. The presentation also covered familiar challenges to long-duration human spaceflight, such as radiation protection and the development of advanced materials. Understanding that this new space venture would need to be a truly national, if not international, venture, DPT suggested cooperating with other Federal agencies, international institutions, academia, and private industry. The team summarized its work as a "focused Agency vision where science becomes the foundation and technology becomes the enabler." In terms of an underlying rationale, the presentation to OMB emphasized that exploration benefits the nation by motivating future generations and promoting international leadership, economic prosperity, and scientific discovery.[125] While DPT had met with OMB in March 2000 and at other times, this December 2000 briefing afforded Isakowitz his first in-depth, formal look at the results of the $5 million seed money he had arranged for decadal planning.

Phase 3: Personnel Changes and a New Name for the Team

On 20 January 2001, President George W. Bush was inaugurated. Decadal Planning Team personnel changes unrelated to the change in administrations took place at this time. Specifically, also in January, Gary Martin took over the leadership of DPT, although Garvin continued on as an active team member. While Garvin was formally trained as a natural scientist, Martin had a bachelor's degree in anthropology, another bachelor's in applied mathematics and physics, and a master's degree in mechanical engineering. Martin came to NASA in 1990 to work as a program manager and Branch Chief in what was then known as Microgravity Sciences and Applications at NASA Headquarters. In 1997, Martin moved to Goddard Space Flight Center and became Chief of the Technology Planning and Integration Office.[126]

The transition from Garvin to Martin reflected a shift from an initial brainstorming phase to an exploratory technology phase. In March 2000, Garvin

---

125. Decadal Planning Team briefing to OMB, NASA Headquarters, 19 December 2000, passim and pp. 2, 20, 24, 34, 38, 44, 68, 70, 86, 88, 94, in key DPT/NEXTBriefings, Lisa Guerra files, NASA HRC.

126. Martin interview, 16 February 2006, pp. 48–49; Glenn Mahone and Bob Jacobs, "NASA's Future Technology Architect Selected," NASA News Release 02-198, 11 October 2002, available at *ftp://ftp.hq.nasa.gov/pub/pao/pressrel/2002/02-198.txt* (accessed 21 August 2008).

assumed the position of Chief Scientist for NASA's Mars robotic exploration program and was increasingly busy with this other portfolio, so he was content to let Martin lead the team. (Martin was officially hired as the Human Exploration and Development of Space [HEDS] Director for Advanced Programs.) Martin and Garvin both described the transition as planned by Rothenberg and Weiler as gradual and orderly. Martin recalls coming aboard the team in the summer of 2000 so he could get up to speed for several months before assuming a leadership role.[127]

In February, Lisa Guerra went on personal leave and Mary DiJoseph, who had a background in space operations, took over Guerra's executive manager duties. While she was on leave, however, Guerra was called back to help brief Courtney Stadd, the new NASA Chief of Staff, and other members of the new administration's transition team about DPT. Guerra formally returned to NASA in August but was reassigned back to the Next Generation Space Telescope project.[128]

The formal phase three of DPT also started in spring 2001 with perhaps the most notable change being its new name: NEXT (NASA Exploration Team).[129] After DPT had initiated studies on nuclear power, in-space propulsion, and astronaut health in space, NEXT continued work on astronaut-robotic integration, systems analyses, advanced concept studies for operations beyond LEO, and other cutting-edge ideas. The steering committee of Rothenberg and Weiler continued to oversee the group's work. Five subteams were established to deal with management, revolutionary technology, astronaut-robotic integration, human health and safety, and outreach. In addition, the groups established a formal Human/Robotic Exploration Science Working Group and named Harley Thronson as its chair.[130]

In mid-2001, other external circumstances prompted shifts in personnel and focus. Like Goldin, Weiler was starting to lose interest in DPT/NEXT because he had more pressing issues with which to contend, particularly the high-profile Mars robotic spacecraft failure. Rothenberg, similarly, could devote little time

---

127. Garvin to Garber and Asner, 26 September 2008; notes from Gary Martin telephone conversation with Steve Garber and Glen Asner, 16 October 2008, NASA HRC.

128. Guerra interview, 26 October 2005, pp. 49–51.

129. A "Management Overview" by Gary Martin and Mary DiJoseph dated 7 May 2001 contains a chart showing possible new names for the team. NEXT was not listed as one of the options. See Space Architect,Doc_Archive,CY01_DPT_PHASE3,LAJOLLA_MEETING, "MANAGEMENT.PDF" file, NASA HRC.

130. Report of the NASA Exploration Team, "Setting a Course for Space Exploration in the 21st Century: June 1999 Through December 2000," 25 February 2002, pp. 8–11, folder 7, Guerra files, NASA HRC. See also NEXT FY02 Annual Report, NASA HRC, for more details.

to DPT because of significant ISS cost problems. Rothenberg retired from NASA in late 2001. Frederick Gregory, an astronaut who had been the head of NASA's safety office, succeeded Rothenberg as the head of the Office of (Human) Space Flight.[131]

ISS cost overruns were causing consternation in Congress and the administration, as well as within NASA and throughout the aerospace community. On 1 November 2001, a subcommittee of the NASA Advisory Council (NAC) chaired by Thomas Young, a highly respected aerospace industry executive and former NASA official, submitted its report on ISS management to the NAC. The "Young Report" was critical of NASA's management of the Space Station.[132]

Dan Goldin resigned as NASA Administrator and left the Agency on 17 November 2001. Goldin had served for almost a year under the new Bush administration, long after most Clinton cabinet secretaries and Agency heads had departed. Goldin was the longest-serving Administrator in NASA's history, having served over nine years during three presidential administrations. He cited various reasons for his resignation decision: personal fatigue, the desire to spend more time with his family after the 11 September 2001 terrorist attacks, and that it simply was time to let somebody else take the reins.[133] At least one reputable news source, however, cited rumors that the new Bush administration had pushed Goldin out, even though it had not found a successor for him.[134] Much has been written about Goldin's positive work as NASA head, as well as his aggressive management style.[135] Regardless, Goldin's visionary approach was the impetus for DPT's start, and thus his departure as Administrator brought with it the possibility of significant changes for NASA and DPT.

---

131. Guerra interview, 26 October 2005, pp. 49–51.

132. See *http://history.nasa.gov/youngrep.pdf* (accessed 25 April 2018) for a copy of this report.

133. See, for example, William Harwood, "NASA Chief To Step Down," *Washington Post* (18 October 2001): A3; Marcia Dunn, "NASA's Longest-Serving Boss Resigning After 10 Years," *Associated Press* (17 October 2001); and Frank Morring, Jr., "Goldin Leaves NASA Post with Agency in a Stall," *Aviation Week & Space Technology* (22 October 2001): 36.

134. See "Goldin Quits Top Space Agency Post, But His Legacy Lingers," *Science* (26 October 2001): 758.

135. See, for example, the news articles cited above, as well as Lambright, "Transforming Government."

4

CHANGE IN LEADERSHIP, CONTINUITY IN IDEAS

SEVERAL NOTABLE TRANSITIONS occurred in late 2001 and 2002 at NASA. In the Administrator's suite, Dan Goldin left and his successor, Sean O'Keefe, came aboard at the end of 2001. Steve Isakowitz, an early proponent of DPT, left OMB and joined O'Keefe at NASA. Rothenberg departed as head of the Office of Space Flight. Fred Gregory succeeded Rothenberg but occupied the position only briefly. Gregory nonetheless remained interested in DPT long after he moved up to the second-highest-ranking position in the Agency, Deputy Administrator. Gary Martin, who enjoyed a strong collegial relationship with Gregory, experienced an increase in stature, if not personal influence, and gained a new job title: Space Architect.

The emphasis of NEXT began to shift, in this context, to reflect the approach and priorities of the new Administrator. The strict embargo that Goldin had imposed on DPT's work was relaxed under O'Keefe, and the membership of the NEXT group also expanded. O'Keefe was interested and supportive of the NEXT group's work, but he was puzzled and disappointed to learn about the "stovepiping" within NASA that NEXT was attempting to overcome. The new Administrator also did not see a clear distinction between science and exploration, which led to a blurring of the "science-driven, technology-enabled" mantra and a weakening of support at the highest level of the Agency for a science-driven agenda.

A New NASA Administrator and New Direction: Early 2002

O'Keefe took over the reins of NASA on 21 December 2001, a month after Dan Goldin's departure. O'Keefe had served at the Pentagon under Secretary of Defense Richard "Dick" Cheney in President George H. W. Bush's administration, first as the Comptroller from 1989 to 1992 and then as the Secretary of the Navy in the administration's final months. He returned to government service in 2001 to serve as Deputy Director of the Office of Management and Budget under George W. Bush. At OMB, he helped formulate the President's Management Agenda, which he often mentioned at NASA. The conventional wisdom (which he did not dispute) was that his budget experience was a strong factor in his selection as NASA Administrator because ISS costs were spiraling out of control.[1]

With strong personal connections in Republican political and national security circles, but no direct experience with space policy or operations, O'Keefe liked to joke self-deprecatingly that his children questioned his ability to lead NASA because he was indeed no "rocket scientist."[2] He did have significant experience with broad space policy issues from his tenure as Secretary of the Navy, Department of Defense (DOD) Comptroller, and deputy head of OMB. Yet O'Keefe was not afraid to admit that he had much to learn about the civilian space program. Shortly after becoming Administrator, he visited most of NASA's Field Centers in an effort to get up to speed quickly.[3]

The DPT/NEXT team formally briefed O'Keefe on 11 February 2002, with Deputy Administrator Dan Mulville, Courtney Stadd, and Ed Weiler in attendance. The briefing package was fairly extensive and discussed such points as NEXT's charter "to create an integrated strategy for science-driven space exploration...not destination-driven," its focus on revolutionary boundary-pushing technology investment priorities for in-space propulsion, nuclear systems, space

---

1. Sean O'Keefe, interview by Asner and Garber, Louisiana State University, Washington, DC, office, 9 February 2006, p. 18, NASA HRC; and various news articles in Sean O'Keefe nomination and confirmation file, file 17612, NASA HRC. For information about the President's Management Agenda, see *http://www.nasa.gov/about/highlights/HP_Management. html* (accessed 27 October 2011). The ISS Management and Cost Evaluation Task Force to the NASA Advisory Council, chaired by A. Thomas Young (*http://history.nasa.gov/youngrep. pdf* [accessed 27 April 2018]), concluded that the ISS program needed major changes in its management and that its cost had almost doubled, going from $17.4 billion to over $30 billion (see pp. 1–5).

2. See electronic file 37809, NASA HRC. O'Keefe also used this same story when he started at NASA. See "Sean O'Keefe Sworn in Amidst the Rockets," *http://www.spaceref.com/news/ viewnews.html?id=425* (accessed 3 September 2008).

3. "Administrator Racks Up Frequent-Flier Miles in Whirlwind Tour," *NASA Headquarters Bulletin* (1 April 2002): 1 (available in electronic file 35728, NASA HRC).

radiation, and optimized astronaut-robotic operations. Interestingly, the last primary chart of the briefing (there were also backup charts) noted explicitly that the NEXT team was not "advocating/developing the next human mission above LEO." According to Martin, the NEXT team was struggling to include Earth science and aeronautics in an overall new strategic vision for NASA and felt that they needed to back off, at least temporarily, on some of the grander space exploration plans.[4] The presentation also included a page on metrics such as science yield, safety, and trip time. As compared to early DPT presentations, this briefing to the new Administrator was more tightly focused but still emphasized a science-based plan of graduated technological steppingstones.[5] The Administrator allowed NEXT to continue and did not ask for any major changes in how it operated.

Although receptive to the ideas that NEXT presented, O'Keefe later revealed that he was surprised that the group considered its proposal for close cooperation of the robotic and human spaceflight camps as novel. As he recollected, "I thought that's what we've been doing for forty years. What's the new thinking on this? This looks like it's what the Agency was commissioned to do, so what you're telling me is this deep, dark secret…?"[6] In O'Keefe's mind, NASA planning was years behind where it should have been. The presentation confirmed for him that the Agency was late in hashing out basic issues concerning its strategic direction. "It was novel in the sense that there was an internal catharsis going on. To an external audience, it was viewed as, 'Isn't that what you do at the office every day?'"[7] Thus, in O'Keefe's mind, the significance of NEXT was not its content, but the fact that the team focused on bringing down barriers to effective communications and planning. O'Keefe's status as an outsider highlighted the perennial divisions between the human and robotic spaceflight camps within NASA. While perhaps greatly overdue, DPT was one effort to bridge the divide.[8]

---

4. Gary Martin, interview by Glen Asner and Stephen Garber, NASA Headquarters, Washington, DC, 16 February 2006, pp. 6–7, NASA HRC.

5. O'Keefe calendar file and Briefing to the Administrator, February 2002, NEXT 2002 folder, Gary Martin files, NASA HRC. The "what we are" and "what we are not" page is p. 13 of this briefing. Compare this briefing, for example, to the DPT presentation dated October 1999, NASA HRC.

6. Sean O'Keefe interview, 9 February 2006, p. 3.

7. Ibid., p. 4.

8. After becoming Administrator, O'Keefe began to realize why NASA Centers are often known as individual "fiefdoms." He tried to change this by implementing a "One NASA" initiative in which employees from Headquarters and all 10 Centers were encouraged to think of themselves as players on a single team.

By early 2002, Martin believed that the team had completed a number of noteworthy and tangible tasks, including establishing a NASA-wide "virtual think tank," developing new strategies and concepts for future human-robotic exploration, developing detailed "roadmaps" identifying technology gaps to be overcome on the way to science-driven exploration, collaborating with existing projects such as the Mars robotic program and the Space Launch Initiative, and identifying areas of cooperation among NASA Enterprises. NEXT continued to promote the concept of allowing scientific questions about life in the universe to dictate exploration activities and destinations. The team also continued to study a wide range of unconventional technological concepts, such as the slinga-tron and blastwave accelerator, as ways to launch payloads into space frequently and inexpensively.[9]

NEXT activities remained embargoed, but the team had expanded con-siderably from when it first began as DPT with approximately 20 people. By March 2002, over 70 civil servants NASA-wide (15 at Headquarters and 60 from the Centers) participated in NEXT activities. The management team grew as well. With NEXT, the Associate Administrators from the Aeronautics, Earth Sciences, and Biological and Physical Research Enterprises joined the Associate Administrators from the Space Science and Space Flight Enterprises on the NASA Strategic Steering Committee, whose five members reported to the NASA Deputy Administrator.[10]

The NEXT team continued to use some of the DPT presentation charts and also continued to use the Lewis and Clark expedition as a historical anteced-ent. Likely realizing the poetic resonance of the upcoming 200th anniversary of Meriwether Lewis and William Clark's famous expedition that began in 1803, team presentations often started by invoking President Thomas Jefferson's personal instructions to Lewis and explaining the resonance to space explora-tion. A typical NEXT presentation started with the quotation from Jefferson to Lewis—"the object of your mission is to explore—and included quotations illustrating "Jefferson's pillars for the country's great enterprise of exploration":

---

9. "NASA Exploration Team Report: Charting America's Course for Exploration & Discovery in the Twenty-First Century," January 2002, p. 3, NEXT 2002 folder, Gary Martin files, NASA HRC. The science questions included the evolution of the solar system, human adaptation to space, Earth's sustainability and habitability, and the possibility of extraterrestrial life. Nearly all of the presentations in this 2002 time period contained these science drivers and ideas for unconventional new technologies.

10. "Proposed NASA Vision and a Strategy for Implementation," NEXT team, 5–6 March 2002, pp. 26, 29, NEXT 2002 folder, Gary Martin files, NASA HRC. For more on NASA's reorganization from "Codes" to "Mission Directorates," please see appendix E.

"Instruments for ascertaining, by celestial observations, the geography of the country" (scientific exploration, enabled by technology)

"…[and to ascertain the suitability of the frontier for] the purpose of commerce" (economic opportunity, enabled by government investment)

"Your observations are to be taken with great pains and accuracy, to be entered distinctly and intelligibly for others as well as yourself" (public engagement, enabled by effective communication)

"You will therefore endeavor to make yourself acquainted, as far as diligent pursuit of your journey shall admit, of…the extent of…[life beyond the frontier]" (the adventure of new discoveries…the *unanticipated*).[11]

U.S. Army Captains Meriwether Lewis and William Clark had led a military expedition called the Corps of Discovery that included 31 other people, the vast majority of whom had been involved with the Corps's initial development and training. After the Louisiana Purchase earlier in 1803, the geographic size of the United States had doubled, but much of this territory was unmapped. Jefferson wanted the Corps to explore this area with an eye toward natural resources that would support western settlement. He supported the Corps with the best logistical supplies (technology) then available and wanted the group to soak up as much information about the new land as possible. As historians at the National Park Service have contended, "By any measure of scientific exploration, the Lewis and Clark expedition was phenomenally successful in terms of accomplishing its stated goals, expanding human knowledge, and spurring further curiosity and wonder about the vast American West."[12] While perhaps the DPT/NEXT comparisons with Lewis and Clark were a bit of a rhetorical

---

11. For a transcript of Jefferson's original, handwritten text, see *http://www.loc.gov/exhibits/ jefferson/168.html* (accessed 10 September 2008). Also see, for example, "NASA's Exploration Team: Products, Evaluations, and Priorities," 30 January 2002, p. 3, NEXT 2002 folder, Gary Martin files, NASA HRC.

12. See *http://www.nps.gov/jeff/historyculture/corps-of-discovery.htm* (accessed 10 September 2008). See also, for example, James Ronda, *Lewis and Clark Among the Indians* (Lincoln, NE: University of Nebraska Press, 2002); and Gary Moulton, ed., *The Journals of the Lewis and Clark Expedition* (Lincoln, NE: University of Nebraska Press, 2001). Thanks to David Kleit for his insights on Lewis and Clark. Another perspective on Lewis and Clark is offered by historian Michael Robinson, who writes about them in his book about 19th-century Arctic exploration and the American character. See *The Coldest Crucible: Arctic Exploration and American Culture* (Chicago: University of Chicago Press, 2006). Robinson concludes his book on p. 164 by quoting President George W. Bush after the Columbia accident proclaiming that "exploration is not an option we choose, it is a desire written within the human heart."

flourish, the analogy was compelling to some participants. The analogy later explicitly made its way into President Bush's 14 January 2004 speech.[13]

In any event, the NEXT members called for investment as soon as possible in key technology areas such as launch vehicles, biomedical countermeasures, and materials. As one of the NEXT presentations put it, "if we do not start now, the country will be constantly a decade away from achieving an integrated vision for space."[14]

More New Leadership in 2002

In the spring of 2002, the human spaceflight program gained a new leader. Fred Gregory took over as acting Associate Administrator for Space Flight when Joe Rothenberg retired from NASA in December 2001, shortly before O'Keefe arrived. O'Keefe officially named Gregory to the Associate Administrator post on 4 March 2002. A former Air Force test pilot and astronaut, Gregory's last job had been as the Associate Administrator for Safety and Mission Assurance.[15]

O'Keefe also named Mary Kicza to be the new Associate Administrator for Biological and Physical Research on 4 March. Kicza moved back to Headquarters from NASA's Goddard Space Flight Center, where she had served in positions of significant responsibility, first as Associate Center Director for Space Science Programs and then as Associate Center Director. An engineer by training, she had also previously served at Headquarters as the Deputy Division Director of the Office of Space Science's Solar System Exploration Division and then as the Assistant Associate Administrator for Technology for the Office of Space Science. Kicza took over for Kathie Olson, who took a position with the White House Office of Science and Technology Policy (OSTP). At Goddard, she also had worked closely with Al Diaz, the Center Director who initially recommended Jim Garvin to lead DPT.[16]

Kicza's appointment was noteworthy for three reasons. As head of the Office of Life and Microgravity Science and Applications (Code U), she would become a stakeholder in August 2002, when she and the Associate Administrators for

---

13. See *https://history.nasa.gov/Bush%20SEP.htm* (accessed 27 April 2018).

14. "NASA's Exploration Team: Products, Evaluations, and Priorities," 30 January 2002, p. 35, NEXT 2002 folder, Gary Martin files, NASA HRC.

15. See *ftp://ftp.hq.nasa.gov/pub/pao/pressrel/2001/01-241.txt* and *ftp://ftp.hq.nasa.gov/pub/pao/pressrel/2002/02-043.txt* for the relevant press releases, and *http://www.jsc.nasa.gov/Bios/htmlbios/gregory-fd.html* and *http://history.nasa.gov/gregory.htm* (all accessed 27 April 2018) for biographical information on Gregory.

16. See *ftp://ftp.hq.nasa.gov/pub/pao/pressrel/2002/02-042.txt* (accessed 27 April 2018) for the relevant press release on Kicza's appointment.

Codes S and M codified their NEXT cooperation in human-robotic space exploration.[17] Second, Kicza had a strong working relationship with Lisa Guerra: Kicza had been Guerra's boss when the latter was tapped to help lead DPT. Once she became head of Code U, Kicza recruited Guerra from the Office of Space Science to work for her as a special assistant for strategic planning. Third, Kicza's entry into this upper level of management also presaged her later appointment to work with the White House on negotiations leading to the Vision for Space Exploration.

In what may have been the most significant hire during his term as NASA Administrator, O'Keefe appointed a former colleague from OMB, Steve Isakowitz, to the position of NASA Comptroller in March 2002. Isakowitz earned bachelor's and master's degrees in aerospace engineering at the Massachusetts Institute of Technology, worked as a project manager and systems engineer in the private sector, and then came to OMB.[18] As the top NASA budget examiner at OMB, Isakowitz provided NASA with formal support and funding to start DPT. He remained interested in DPT as it progressed and evolved into NEXT and as the leadership of the federal government shifted from Democratic to Republican control. No one else outside of NASA had as much knowledge of the DPT/NEXT activities. A trusted advisor to O'Keefe at OMB, Isakowitz retained O'Keefe's confidence at NASA. Because of Isakowitz's technical knowledge of aerospace issues and his financial acumen, the Administrator turned to Isakowitz for analytical support when he needed to make tough decisions. Isakowitz would play a key role in later negotiations with the White House over the scope and content of what would become the VSE.

In April 2002, Administrator O'Keefe gave a speech at Syracuse University in which he discussed his vision for NASA, particularly his thoughts on the role of science at the Agency. Reflecting the tenor of the DPT/NEXT presentations, he said that "NASA's mission…must be driven by the science, not by destination…science must be the preeminent factor."[19] O'Keefe also emphasized that NASA "will let the science of exploration and discovery tell us where to go next" and that "there is a necessary link and connection between our human space flight program and our work in robotics. NASA must eliminate the

---

17. "NEXT FY02 Annual Report," pp. 2–3, Harley Thronson files, NASA HRC.

18. No press release was issued for this new appointment. O'Keefe's calendar notes that he was scheduled to give remarks at Isakowitz's farewell from OMB party on 12 March 2002.

19. NASA Facts, "Pioneering the Future" (address by the Honorable Sean O'Keefe, NASA Administrator, to the Maxwell School of Citizenship and Public Affairs, Syracuse University, 12 April 2002), p. 4, available at *https://historydms.hq.nasa.gov/sites/default/files/DMS/e000040280.pdf* (accessed 8 August 2018).

stovepipes and build an integrated strategy that links human space flight and robotic space flight in a steppingstones approach to exploration and discovery."[20]

O'Keefe recalled a few years later that he gave this speech to reassure the concerned scientific community about its prominent role in NASA's planning deliberations. While he felt that he could not pledge money for specific programs at that time, O'Keefe hoped to gain the confidence of the scientific community with general reassurances that he supported the Agency's scientific programs and aspirations. He did not mean to imply that science should trump human exploration by astronauts but rather that they should go hand in hand.[21]

In August 2002, the Associate Administrators for Space Science (Weiler), Biological and Physical Research (Kicza), and Space Flight (Gregory) signed a Memorandum of Agreement (MOA) formally establishing the NEXT team.[22] As noted above, these offices were known as Codes S, U, and M, respectively; hence the SUM nickname. The Associate Administrators for Earth Science and Aeronautics who had been part of the informal NEXT steering committee a few months earlier did not sign the agreement.

Fred Gregory's term as head of the human spaceflight organization lasted less than a year, until 12 August 2002, when O'Keefe appointed him Deputy Administrator. In his new role, Gregory gained responsibility for overseeing long-range planning. To consolidate the activities of various groups that may have been engaged in long-range strategic planning efforts across the Centers, Gregory recommended the creation of a new position with broad responsibilities for coordinating long-term planning. He suggested that O'Keefe name the head of NEXT, Gary Martin, as the Agency's first "Space Architect." Seeking greater coordination across NASA for a wide range of activities, O'Keefe accepted Gregory's suggestion. In October 2002, Martin continued working for Gregory, formally reporting to the Deputy Administrator as the Agency's first Space Architect.[23] In this vein, O'Keefe viewed Gary Martin's job as leader of NEXT, and later as Space Architect, as that of a "convener and coordinator" of disparate NASA activities and planning functions.[24]

As Space Architect, Martin gained responsibility for leading the development of the Agency's space strategy and for reporting to NASA's Joint Strategic

---

20. O'Keefe speech at Syracuse University, p. 9.
21. O'Keefe interview, pp. 6–8. On p. 6, he also claims that he does not see a "real distinction between exploration and scientific objectives."
22. "NEXT FY02 Annual Report" (undated), p. 2, folder 13, NEXT Briefings/Annual Reports, file 12, Lisa Guerra files, NASA HRC.
23. See Gary Martin interview, 16 February 2006, pp. 11–12; and 11 October 2002 press release (*ftp://ftp.hq.nasa.gov/pub/pao/pressrel/2002/02-198.txt*).
24. O'Keefe interview, pp. 9–12. The quotation is from p. 12.

Assessment Committee (JSAC). The JSAC was one half of an attempt to streamline the management and planning of new space initiatives to create an integrated space plan; the other half was the Space Architect and his team.[25] The Office of Space Science was to lead an effort reporting to Martin to develop a rationale for exploration beyond LEO. Also in the summer 2002 timeframe, Gregory directed NASA to conduct a study with four objectives: develop a rationale for exploration beyond LEO, develop roadmaps for steppingstones to put astronauts on Mars, make design reference missions the basis for these roadmaps, and provide practical recommendations

Gary Martin, NEXT chair. (NASA)

on what to do immediately to accomplish these goals. This 90-day study (not to be confused with Aaron Cohen's, discussed earlier) was conducted Agency-wide but led by JSC and managed at Headquarters by Gary Martin during September–November 2002. It became known as the *Exploration Blueprint Data Book* (or simply "blueprint study").[26] The JSAC would continue via study teams that immediately followed the VSE announcement in January 2004.[27]

---

25. See Gary Martin, "New Initiative Process for Space: Joint Strategic Assessment Committee (JSAC)," 4 March 2002, NASA HRC. Interestingly, the date of the presentation was the same as when Gregory was named head of NASA's human spaceflight organization, but still several months before Martin was named the Space Architect.

26. See Bret G. Drake, *Exploration Blueprint Data Book* (NASA TM-2007-214763). For some reason, this report was "published" in 2007. The foreword (p. 1) explains some of the time sequencing before this TM publication begins in earnest with the report from November 2002. A copy of this TM is in the Gary Martin files. Also see "Exploration Blueprint Input: Integrated Space Plan," 21 November 2002, Space Architect 2002 folder, Gary Martin files, NASA HRC. This presentation also offers an interesting chart on p. 6 showing how the Space Architect would focus on requirements and systems engineering distilled from the Space Act and NASA's Strategic Plan, as well as scientific drivers, to produce gap analyses that could then produce the integrated space plan.

27. See John Campbell, "Red Team Brief to JSAC," 25 March 2004, NASA HRC. See also Gary Martin interview, 16 February 2006, pp. 13–14, for a brief mention of the JSAC and its origins.

NEXT FY01 Annual Report cover. (NASA from Trish Pengra-electronicfiles\DPT.NEXT\Annual Report\AR)

Gregory and O'Keefe expected Martin to draw on his background in systems analysis to identify and assess the long-term investments needed to bring this strategy to fruition. As Space Architect, he also was expected to lead the ongoing effort to identify a robust rationale for NASA's space strategy, balancing such factors as scientific goals, commercial development, and educational benefits.[28] On paper, Martin served in two distinct roles: as Space Architect and as leader of the NEXT team. In practice, these two positions were nearly indistinguishable.

Shortly after his appointment as Space Architect, Martin gave a presentation about NEXT at Kennedy Space Center. By then, NEXT had broadened its membership to include an Agency-wide team of approximately 50 people. In his presentation, Martin highlighted three of NEXT's new technology investment priorities: in-space propulsion (a funded FY 2002 initiative), a nuclear systems initiative (included in the President's FY 2003 budget), and a space radiation program (also included in the President's FY 2003 budget).[29]

28. Gary Martin, "NASA's Integrated Space Plan," p. 4.
29. Gary Martin, NEXT presentation (presented at the Cape Canaveral Spaceport Symposium, October 2002), Space Architect 2002 folder, NASA HRC.

Contrasting his team's new exploration strategy with that of the Apollo program, Martin called for the DPT-like steppingstones approach instead of the "giant leap" forward approach of the 1960s. The historical context for the Apollo program was markedly different from the environment in the early 21st century. Martin noted that the Apollo program had focused exclusively on sending humans to the Moon as an end in itself, with a presidentially mandated deadline and nearly unrestricted spending authority. The new paradigm, however, needed to be flexible enough to support sending humans and robots to different destinations with scientific discovery as the driver, with a longer timeframe but vastly fewer financial resources. Invoking O'Keefe's "One NASA" management theme, Martin explained that all NASA elements would need to pool their resources and considerable skills to achieve these goals. While limited resources would allow only a gradual, evolutionary approach, this new exploration strategy would be revolutionary in the sense that it would require great creativity and numerous technological breakthroughs to achieve.[30] Utilizing architectural trade studies, programmatic and technology roadmaps, and gap analyses, Martin's Space Architect team was tasked with producing an integrated space strategy and prioritized technology initiatives.[31]

The formal Space Architecture Team included members from NASA's programmatic Enterprises, its functional support offices, and the various Field Centers. Several of these people—such as Harley Thronson, Lisa Guerra, John Mankins, Lynn Harper, and Rud Moe—had participated in DPT.[32] Doug Comstock, who worked for Steve Isakowitz at OMB and then moved over with him to the NASA Comptroller's office, also was part of the team. Doug Cooke, who worked on human exploration issues for many years at JSC and later became the Agency's Deputy Associate Administrator for Exploration Systems, worked closely with the team.[33]

---

30. Gary Martin, "NASA's Integrated Space Plan," pp. 2–3.

31. Ibid., p. 6.

32. John Mankins worked in aerospace advanced concepts and served as the Chief Technologist for the HEDS Enterprise. See *http://www.spaceref.com/news/viewsr.html?pid=18166* (accessed 5 September 2008). An engineer by training, Rud Moe served as the servicing missions manager for the Hubble Space Telescope. Lynn Harper worked at NASA Ames Research Center in the field of astrobiology.

33. See, for example, Gary Martin, "Integrated Space Plan Development: Introduction to Joint Strategic Assessment Committee" presentation, 13 December 2002, pp. 7, 14.

Space Policy Ruminations in the White House

Earlier in the year, in April 2002, Gil Klinger began his new job as director of space policy for the National Security Council. Like O'Keefe, Klinger had little prior work experience with civilian space issues. He had about a decade of experience working on military and intelligence space issues, however, and had recently completed a stint as head of the National Reconnaissance Office's policy organization. Klinger became interested in space as a young boy during the 1960s, in the heydays of the Apollo program. In the mid-1990s, while working in the defense and national security communities, he began to deal with national space policy issues that involved NASA. In the mid- to late 1990s, he began to feel that NASA was adrift.[34] Klinger believed that NASA needed the support of the White House, particularly the President, to articulate a compelling mission and to reengage the public to counter years of disinterest and ambivalence about the Agency's mission. Yet he also believed that the President was unlikely to approve a major restructuring of NASA's direction any time soon, given the consuming nature of other higher-priority issues, particularly national security.

The fall of 2001 presented no opportunities for discussion of civilian space policy at the White House. In the immediate aftermath of the 11 September 2001 terrorist attacks, the President and his staff were consumed with several major security initiatives, including an active war in Afghanistan, the broader war on terrorism, planning for the invasion of Iraq, and massive reorganizations of the national intelligence agencies and the creation of the Department of Homeland Security. Meanwhile, on the personnel front, Dr. John Marburger became the permanent presidential Science Advisor on 29 October 2001, succeeding two previous temporary advisors to President Bush. During the span of several years at OSTP, the assistant director for space and aeronautics changed several times. At NASA, Administrator Dan Goldin resigned effective 17 November 2001.[35]

---

34. Gil Klinger, interview by Asner and Garber, NASA Headquarters, Washington, DC, 11 January 2006, pp. 6–7, NASA HRC.

35. See *http://web.archive.org/web/20080517145017/http://www.ostp.gov/cs/about_ostp/previous_science_advisors* (accessed 27 October 2011), *http://history.nasa.gov/prsnnl.htm* (accessed 27 April 2018), and Brett Alexander, interview by Asner and Garber, NASA Headquarters, Washington, DC, 19 January 2006, p. 3, NASA HRC. Scott Pace took over from Vic Vilhard as the OSTP assistant director for space and aeronautics after Vic Vilhard left; once Pace went to NASA to be the Deputy Chief of Staff for O'Keefe, Bill Jeffrey took over before moving to the National Institute of Standards and Technology (NIST).

Klinger saw few signs of White House interest in civilian space activities in the spring of 2002. Observers in the space community assumed that President George W. Bush would be hesitant to repeat a perceived failure of his father, whose Space Exploration Initiative collapsed under the burden of high projected costs and the lack of a compelling rationale. Like Isakowitz several years earlier, Klinger encountered resistance within the administration to supporting NASA's human spaceflight efforts.[36]

Nevertheless, Klinger quickly developed a strong working relationship with his civilian space policy colleague in the Office of Science and Technology Policy, Bretton "Brett" Alexander. An aerospace engineer, Alexander had worked in Russia and in the commercial sector in the United States before taking a government job in the Federal Aviation Administration's Office of Commercial Space Transportation and then moving over to OSTP as a space staff member. Alexander and Klinger began to consider the broad parameters of NASA's mission and to weigh options for new space policies. They hoped for a reinvigoration of civilian space activities, particularly in the realm of human spaceflight.

While the upper levels of the White House were not focused on civilian space policy, in June 2002, the President formally authorized his NSC staff to conduct a review of national space policies. Stephen Hadley delegated this responsibility to Frank Miller, who in turn made Klinger the key staff lead. Alexander took an active role in the effort. The first two topics Klinger and Alexander covered were commercial remote sensing and space transportation. Their charge was to review and update all other national space policies as they saw fit. These space policy directives were completed at an uneven pace over the next several years.[37]

During 2002, Alexander and Klinger had been thinking informally about national space policy, anticipating the centennial-of-flight anniversary in December 2003. This high-profile anniversary seemed like a propitious time

---

36. Klinger interview, 11 January 2006, pp. 2–5. He also related a colloquial anecdote demonstrating unsatisfying circular logic: asking lay people why NASA had a Space Shuttle (supposed answer: to get to the International Space Station) and why NASA maintained the ISS (supposed answer: so the Shuttle would have someplace to go). See p. 3.

37. Alexander interview, p. 3. The space transportation policy directive was almost ready for signature at the end of January 2003, but then the Columbia Space Shuttle accident on 1 February 2003 delayed its release by almost two years; it was released on 6 January 2005. (The Columbia accident and its impact on the formulation of the VSE is the subject of the next chapter.) The commercial remote sensing policy was released on 25 April 2003. Other policies on Earth science, global positioning, aeronautics, and national space policy in general were released in 2004–2006. These various policies may be found at *http://web.archive.org/web/20080514201001/http://www.ostp.gov/cs/issues/space_aeronautics* (accessed 27 October 2011).

to unveil a major new policy initiative, but the event remained too far off in the future for senior White House officials to give it any attention.[38]

Another Blue-Ribbon Commission Calls for Change

In November 2002, the Commission on the Future of the Aerospace Industry issued a report calling for a new national aerospace policy. Former Congressman Robert Walker chaired the congressionally chartered commission that had half its members appointed by the White House and had a charter covering both space and aeronautics. Eleven other prominent individuals from different backgrounds served on the Commission, including astronaut Buzz Aldrin; astronomer Neil deGrasse Tyson; and chairman, chief executive officer, and general manager of the Aerospace Industries Association John Douglass.[39] As the NASA Space Architect, Gary Martin provided information to the Commission.

The Commission addressed several topics relevant to future space exploration in the course of its work. In June 2002, for example, the Commission issued its third interim report, which called for the "U.S. aerospace industry to create superior technology…to reach for the stars" and maintain the U.S. space launch infrastructure. It also noted the need to retain solid rocket motor production capability in light of the Space Shuttle's anticipated retirement and other launch customers moving to liquid fuels.[40]

The Commission's final report decried the federal government's failure to provide adequate funding for aerospace research and infrastructure development. It also expressed concern about the aging of the aerospace industry workforce. The report's "vision" was "anyone, anything, anywhere, anytime," and it called for a Government-wide framework to implement a national aerospace policy that would remove regulatory barriers and increase investment in this sector, facilitating U.S. aerospace leadership globally. Some critics dismissed as

---

38. See, for example, Klinger interview, p. 5.

39. A copy of the report is available at *http://history.nasa.gov/AeroCommissionFinalReport.pdf* (accessed 27 April 2018). A hard copy of the report, its executive summary, as well as two short analyses of it, are available in file 18065, NASA Historical Reference Collection. The analyses are Randy Barrett, "Space Commission Report Receives Mixed Reviews," *Space News* (25 November 2002): 16; and a forwarded lengthy email from P. Graham O'Neil, United Space Alliance, to Roger Launius, 20 November 2002.

40. See *http://www.hq.nasa.gov/office/hqlibrary/documents/o51233742.pdf* (accessed 10 September 2008), pp. 1, 19. The report also recommended that the Secretary of Defense ask the Defense Science Board to review the Pentagon's policy on potential aerospace industrial base consolidation. This recommendation was heeded, and the *Report of the Defense Science Board/ Air Force Scientific Advisory Board Joint Task Force on Acquisition of National Security Space Programs* was issued in May 2003. (See *http://www.globalsecurity.org/space/library/report/2003/ space.pdf* [accessed 10 September 2008]).

unrealistic the few specific proposals that were included in the final report, such as creating an Office of Aerospace Development in every federal department and establishing a Bureau of Aerospace Management within OMB.[41]

By late 2002, therefore, scattered groups within the executive branch and throughout the aerospace community had given serious consideration to the future of NASA. The DPT team completed its initial brainstorming phase and transitioned smoothly to NEXT, engaging individuals from various disciplines and programs throughout NASA. As noted at the end of chapter 3 and earlier in this chapter, the transition from DPT to NEXT began informally in early 2001 and was formalized with an MOA in August 2002. Administrator O'Keefe supported these efforts and began incorporating NEXT ideas into his speeches and formal Agency plans. At the White House, Brett Alexander and Gil Klinger were a step ahead of their bosses, informally sharing ideas for a new civilian space policy. Concerned about the overall decline of the U.S. aerospace industry, even Congress supported efforts to rethink how NASA operated. None of them, however, could anticipate the challenges that NASA would soon face.

---

41. As examples, Barrett notes that Walker conceded that creating a new bureau within OMB would be "one of the more difficult pieces" and quotes Michael Bevin of the American Institute of Aeronautics and Astronautics (AIAA) that "It's one thing to make a recommendation but another thing to implement it." *U.S. Aerospace Industry: Progress in Implementing Aerospace Commission Recommendations, and Remaining Challenges* (Washington, DC: U.S. Government Accountability Office Report GAO-06-920, September 2006), pp. 9, 16–18, 34–35.

5

THE COLUMBIA ACCIDENT AND ITS AFTERMATH

AT APPROXIMATELY 9:00 a.m. eastern time on 1 February 2003, during mission Space Transportation System (STS)–107, the Columbia orbiter disintegrated in midair approximately 15 minutes before its scheduled landing at Kennedy Space Center (KSC), killing all seven crew members. As investigators soon learned, a chunk of insulating foam had come off the Space Shuttle's cryogenic external fuel tank and punched a hole in the left wing of the orbiter vehicle's thin skin on ascent. On descent, the hole permitted superheated plasma gas to burn through a wing structure, causing the vehicle to lose control, and then the orbiter vehicle came apart after experiencing loads it could not sustain. After overcoming their initial shock, NASA employees moved swiftly to implement the Agency's contingency plans and to begin debris recovery operations with the assistance of scores of volunteers and experts from other government agencies. Within 2 hours, Sean O'Keefe activated a mishap investigation board, which later became known as the Columbia Accident Investigation Board (CAIB).[1]

This was the second major accident in 22 years of Shuttle operations—the first accident occurred when the Challenger orbiter broke apart during launch in 1986. To some observers, the Columbia accident confirmed the deficiencies of the aging Shuttle fleet. In at least one way, the Columbia accident was even more disturbing than the Challenger accident because the former highlighted the inherent fragility of the Shuttle's basic design configuration—shedding foam could not be truly eliminated from future missions. Beyond the technical

---

1. Information about the Columbia accident and the CAIB is available at *http://history.nasa.gov/columbia/ CAIB_reportindex.html.*

October 2001 portrait of the STS-107 crew. Seated in front are (left to right) Rick D. Husband, mission commander; Kalpana Chawla, mission specialist; and William C. McCool, pilot. Standing are (left to right) David M. Brown, Laurel B. Clark, and Michael P. Anderson, all mission specialists; and Ilan Ramon, payload specialist representing the Israeli Space Agency. (NASA sts107-s-002)

problems, the CAIB Report noted that "NASA's organizational culture had as much to do with this accident as foam did."[2] Financial, technical, and political hurdles delayed efforts that began in the mid-1980s to develop a new vehicle to replace the Shuttle, a technology that had its origins in designs from the 1960s and early 1970s. The latest iteration of the Shuttle replacement vehicle was being designed at the time of the accident. NASA's leadership and many

---

2.   Columbia Accident Investigation Board, *Report* (hereafter CAIB Report), vol. I (Washington, DC: NASA, 2003) p. 12. Chapter 7 of the CAIB Report is titled "The Accident's Organizational Causes." On p. 195, at the beginning of chapter 8, "History as Cause: *Columbia* and *Challenger*," the CAIB states that "[t]he foam debris hit was not the single cause of the *Columbia* accident, just as the failure of the joint seal that permitted O-ring erosion was not the single cause of *Challenger*. Both *Columbia* and *Challenger* were lost also because of the failure of NASA's organizational system." On p. 201, the CAIB Report states that the "organizational structure and hierarchy blocked effective communication of technical problems. Signals were overlooked, people were silenced, and useful information and dissenting views on technical issues did not surface at higher levels."

In a hangar at KSC, pieces of debris from Columbia lie across the grid on the floor on 28 February 2003. Members of the Columbia Restoration Project Team examine pieces and reconstruct the orbiter as part of the investigation into the accident. (NASA KSC-03pd0524)

people throughout the space community believed that a new transportation system was long overdue.[3]

Equally important, the first half of 2003 was a period of heightened anxiety and uncertainty for most Americans. The 11 September 2001 terrorist attacks had occurred less than 18 months earlier, and the U.S. military campaign against Al-Qaeda and affiliated terrorist groups in Afghanistan and elsewhere was still in an early phase. Just four days before the Columbia accident, President Bush indicated in a State of the Union Address his serious intent to invade Iraq if Saddam Hussein did not destroy his nation's supposed stockpiles of weapons of mass destruction. Secretary of State Colin Powell addressed the United Nations Security Council just four days after the accident, captivating the world with a dramatic speech claiming links between Saddam Hussein and terrorists; the existence of massive caches of nuclear, biological, and chemical weapons; and the intention of Iraq to inflict harm on the United States. Although NASA quickly dispelled concerns that the Columbia astronauts were victims of terrorism, echoes of the 11 September 2001 attacks reverberated prominently

---

3. CAIB Report, vol. I, pp. 110–119. See also John Logsdon, "'A Failure of National Leadership': Why No Replacement for the Shuttle?" in *Critical Issues in the History of Spaceflight*, ed. Steven J. Dick and Roger D. Launius (Washington, DC: NASA SP-2006-4702, 2006), pp. 269–300.

President George W. Bush speaks at the Columbia memorial service at JSC on 4 February 2003. (NASA jsc2003e05893)

on the day of the accident and continued to influence the policy context in subsequent months.[4]

For NASA and the White House, merely returning the Shuttle to flight would have been an inadequate response to the accident, given the condition and age of the Shuttle fleet and the desire of the nation to retain its reputation as the global leader in technology and human spaceflight. Short of abandoning the Shuttle and International Space Station entirely, NASA's best option for preventing accidents over the long term seemed to be to build a new, safer space transportation system.

NASA's leadership had not anticipated the occurrence of a catastrophic event that would allow for a complete reconsideration of the Agency's mission. Assuming that opportunities for overhauling the mission of NASA lay far in the future, the NEXT/DPT teams conducted their work in relative obscurity and with no great urgency prior to the accident. The teams focused on generating new ideas and vetting options for space exploration rather than on developing

---

4. CAIB Report, vol. I, pp. 85, 93–94. The transcript of Bush's 28 January 2003 State of the Union speech is available at *http://www.presidency.ucsb.edu/ws/index.php?pid=29645&st=&st1=* (accessed 02 September 2009), and the transcript of Powell's 5 February 2003 speech before the United Nations is available at *https://2001-2009.state.gov/secretary/former/powell/remarks/2003/17300.htm* (accessed 1 June 2017).

a single, coherent policy that could be implemented at the nearest opportunity. In a sense, the group was still "sneaking up on Mars" when the Columbia accident motivated officials at the highest levels of the executive branch to support a dramatic policy shift. While they mourned the loss of astronauts who had risked their lives for human spaceflight, a handful of officials on O'Keefe's leadership team could not help but wonder what the accident meant for the Agency, the Shuttle Program, and the future of human spaceflight. The accident forced many people, especially those involved in NASA's long-term exploration planning efforts, to reexamine their assumptions. The Columbia accident provoked a sense of urgency among those at NASA who had been strategizing for years on how to gain support for a bold, new initiative from those at the White House who had the power to implement a new exploration plan.

A White House Divided

The leadership of NASA and supporters of human spaceflight in the Executive Office of the President worried as much about the debate over human spaceflight within the Bush administration as they did about public opinion. O'Keefe viewed the period as a perilous time for NASA. He believed that the entire civilian space program would be in jeopardy if NASA failed to identify clear objectives for itself and enlist the support of policy-makers and White House officials. O'Keefe was not alone in believing that NASA might be disbanded, with various component programs distributed to other federal departments or agencies, such as DOD, the Federal Aviation Administration (FAA), and the National Oceanic and Atmospheric Administration (NOAA).[5] Klinger believed that "we were at risk of losing the entire program right then and there," meaning the whole human spaceflight effort, which most people viewed as NASA's raison d'être. According to Klinger, "a fairly vigorous debate [emerged] within the White House about whether or not we should simply have only a robotic program." In particular, senior OSTP officials to whom Brett Alexander reported expressed great skepticism about the value of human spaceflight.[6]

On the morning of the Columbia accident, a collection of top- and mid-level officials gathered in the White House situation room. Alexander and Klinger, along with Joseph Wood, an Air Force Colonel on loan to Vice President Cheney's staff, were among the few people in the situation room with space policy experience or knowledge. Most senior officials, such as then–National Security Advisor Condoleezza Rice and Homeland Security Advisor Tom

---

5. O'Keefe interview, pp. 14–17, 21.
6. Klinger interview, p. 9.

Ridge, had little or no experience with NASA or space policy. Alexander, Klinger, and Wood spent most of the day trying to learn the specifics of the accident and answering general questions from top government leaders. They occasionally stopped to watch televised NASA briefings. No NASA officials were at the White House initially.[7]

Officials in the White House without a direct role in recovery operations quickly focused their attention on questions about the future of the Agency and human spaceflight. Within a day or so, a loose group called the Columbia Accident Coordinating Group began meeting. White House cabinet secretary Brian Montgomery, who had just started his job as deputy assistant to the President and cabinet secretary in January 2003,[8] convened this group of internal White House staffers. Alexander, Klinger, and Wood were the first individuals to join the group.[9] They agreed from the start that human spaceflight should continue and that, at the earliest opportunity, administration officials should emphasize publicly their intentions to maintain a human presence in space. A clear statement of support for human spaceflight early on, they believed, would help maintain momentum for returning the Shuttle to regular operations.

Although the debate continued for nearly two months, the President and Vice President pledged in the days after the accident to continue the human spaceflight program. At a memorial service at Johnson Space Center on 4 February, President Bush declared that the nation would return to flying people in space. Vice President Cheney echoed this sentiment at a memorial service in Washington on 6 February. Brett Alexander saw these speeches as touchstones and carried around copies of them to remind his superiors, when they would express doubt about the value of continuing the program, of the President's support for human spaceflight. Klinger similarly viewed these public statements as ironclad and necessary support for human spaceflight in the face of opposition from others in the White House.[10]

---

7. See, for example, Alexander interview, pp. 5–7.

8. Before taking on this position as cabinet secretary, Montgomery had worked for two years as the head of the President's advance team (for trips). He left the White House for a position at the Department of Housing and Urban Development in 2005. As cabinet secretary, he coordinated the flow of information and policy among the cabinet members and relevant White House staff for the President. See *http://web.archive.org/web/20070922100527/http://www.hud.gov/offices/hsg/bios/bmontgomery.cfm* (accessed 27 October 2011).

9. Others, such as Dave Radzanowski from OMB, joined the group later. The coordinating group met for several months, until the CAIB issued its report in August. See, for example, Alexander interview, pp. 7–8.

10. See Alexander interview, pp. 12–13; Klinger interview, pp. 8–10; and Isakowitz interview, p. 21. The speeches are available at *http://history.nasa.gov/columbia/executivebranch.html*. On the day of the accident, Bush said that "the cause in which they died will continue…. Our journey into

Having considered options for redirecting the civilian space program before 1 February, Alexander was especially keen to engage in policy discussions in the aftermath of the Columbia accident. He gained permission from Brian Montgomery to pull together a small group of White House staff members involved in space policy for informal discussions. The team became known as the Splinter Group, with some of the same people as on the Columbia Accident Coordinating Group.[11]

The core group initially consisted of Alexander, Klinger, and Wood. It later expanded to include OMB staffers such as Dave Radzanowski, Amy Kaminski, Jim Martin, and Paul Shawcross. By design, the Splinter Group did not include any NASA representatives. Alexander and Montgomery sought to keep the group small and informal to allow for general discussions and brainstorming sessions. Over the course of several months in the spring of 2003, the group debated various rationales for space exploration, as well as the potential applicability of historical analogies for the space program, such as the Lewis and Clark expedition. Montgomery supported the Splinter Group but did not ask the group to develop any particular policy or plan.[12]

In May 2003, the Splinter Group produced a white paper intended for circulation only within the White House. This white paper made several notable points. First, as the CAIB Report would point out publicly several months later, the white paper decried the lack of a clearly articulated, "compelling vision" since the collapse of the Soviet Union, if not the end of the Apollo program. It also noted that U.S. national security space programs had been beset with budgetary and schedule problems. The white paper called for bold presidential leadership to reverse the situation and initiate a new mission for NASA. It also called for building key technologies in sequential fashion: replacing the Shuttle by 2010 with a new, dramatically safer spacecraft; developing a next-generation launcher by 2020 that would significantly reduce the cost of accessing space; and building new in-space transportation capabilities such as space

---

space will go on." See *http://www.presidency.ucsb.edu/ws/index.php?pid=181&st=space&st1=* (accessed 2 September 2009). A few days later, at a 6 February memorial service for the Columbia crew, Cheney eulogized the crew thus: "[W]hile many memorials will be built to honor Columbia's crew, their greatest memorial will be a vibrant space program with new missions carried out by a new generation of brave explorers." See *http://georgewbush-whitehouse. archives.gov/news/releases/2003/02/20030206-8.html* (accessed 2 September 2009). A few months later, on 1 June 2003, Bush and Russian President Vladimir Putin issued a brief joint statement in which they affirmed their nations' commitments to returning the Shuttle to flight and to continuing to assemble the ISS. See *http://www.presidency.ucsb.edu/ws/index. php?pid=63302&st=putin&st1=space* (accessed 2 September 2009).

11. Alexander interview, pp. 13–14.

12. Ibid., pp. 13–15.

nuclear propulsion. Moreover, it called for renewed robotic efforts to search for extraterrestrial life. Finally, it called for the NASA Administrator's position to be elevated to formal cabinet rank.[13]

Alexander sent this white paper up to Richard Russell, the OSTP Associate Director for Technology with responsibility for space and aeronautics.[14] When Alexander and William Jeffrey, the OSTP assistant director for space and aeronautics, presented the paper to Russell, he skimmed through it while they waited. According to Alexander, and for reasons that are not entirely clear, Russell groused, "this is exactly what we *don't* need."[15] Russell may have opposed the continuation of human spaceflight or merely worried that the tone of the paper was too aggressive. In any event, the policy development process within the White House then entered a brief lull until late June 2003.

The Accident and NASA

Key members of NASA's senior staff also began meeting on an informal basis just after the Columbia accident to consider options for the Agency's future. Isakowitz, the NASA Comptroller and confidant of O'Keefe, approached Gary Martin not long after the accident. Isakowitz asked Martin if he had thought seriously about sending humans back to the Moon. As Martin recalls, Isakowitz said, "Look, we really ought to have something in our back pocket…. What are the first steps?" Martin agreed that the Moon was a logical first step for humans to go beyond LEO, but he had little of value to offer Isakowitz. The DPT/NEXT teams had not conducted formal lunar studies, as they had focused on identifying key capabilities to reach multiple destinations rather than any single target. Martin's general feeling was that DPT/NEXT had focused on Mars, as well as libration points. Both Isakowitz and Martin agreed that the time was right to evaluate lunar options.[16]

Martin subsequently asked space architecture teams from JSC and LaRC to prepare studies on returning astronauts to the Moon. The LaRC team presented its plan for "Modular Systems for Exploration" on 13 March 2003 to Martin and his Space Architect team. They took as their charge to send two to

---

13. "The Next Space Age," white paper, May 2003, Alexander file, pp. 1–5. The initial quotation is from p. 2.

14. See *https://georgewbush–whitehouse.archives.gov/government/russell-bio.html* (accessed 12 June 2018) for biographical information about Russell.

15. Alexander interview, p. 15. Biographical information about Jeffrey is available at *http://pubs. acs.org/cen/news/83/i22/8322egovc1a.html* (accessed 27 October 2011), for example.

16. Isakowitz interview, p. 27; Martin interview, p. 14. The quotation is from the Martin interview, p. 14.

four astronauts to the lunar South Pole for three days (total trip time would be a week), in part to search for in situ resources. The study included focused and broad approaches; the former would utilize existing modular technologies to keep costs down, and the latter would support other efforts, such as military and commercial satellites, as well as human voyages to libration points and Mars.[17]

Because of the accident, Isakowitz worried that NASA was at the "edge of losing the whole human spaceflight program."[18] While concerned that NASA might appear greedy or opportunistic in asking for a major new program soon after the deaths of seven astronauts, Isakowitz also felt that the Agency was in serious jeopardy, as O'Keefe heavily emphasized in retrospect.[19]

The Columbia Shuttle accident was the second that resulted in the loss of all crewmembers, and some space policy insiders were concerned that public perception might be that NASA had not changed its safety culture since returning to flight after the Challenger accident. In addition, many insiders believed that many members of the public failed to understand the extreme risks and inherent difficulty of human spaceflight. Also, the accident pointed out how technically fragile the Space Transportation System was—increased monitoring of foam strikes to the orbiter was not sufficient to prevent accidents because NASA still did not have a demonstrated capability to repair orbiters in orbit. The Shuttle had flown over 100 times, yet NASA was reduced to again considering it an experimental vehicle.[20] For these reasons, NASA's leadership worried that the accident could lead to the termination of the Shuttle Program and, ultimately, the loss of the entire human spaceflight program.

After working on exploration scenarios with Martin and preliminary cost estimates with specialists on his own staff, Isakowitz decided to seek O'Keefe's approval to start moving ahead with a bigger plan. He raised the issue within the context of the annual Agency budget discussion in February and March,

---

17. Space Architect Architecture Team, "Modular Systems for Exploration" presentation, 13 March 2003, see especially pp. 3, 4, 11, 41. This document was provided by Patrick Troutman, who apparently led the LaRC team, and is in the Gary Martin files, NASA HRC. See also email from Martin to authors, 9 July 2009, Printed Emails to Glen Asner and Steve Garber folder, NASA HRC.

18. Isakowitz interview, p. 27.

19. O'Keefe interview, pp. 14–17, 21.

20. The STS-5 mission, which launched on 11 November 1982, was considered the first operational Shuttle mission (the first four Shuttle missions were considered experimental). For a good discussion of this point, see "The Shuttle Becomes 'Operational'," section 1.4 of the CAIB Report, volume 1, August 2003, pp. 23–24, available at *https://history.nasa.gov/columbia/Troxell/Columbia%20Web%20Site/CAIB/CAIB%20Website/CAIB%20Report/Volume%201/Part%201/chapter1.pdf* (accessed 28 June 2018) and *https://www.nasa.gov/mission_pages/shuttle/shuttlemissions/archives/sts-5.html* (accessed 26 June 2018). Thanks to Colin Fries for his help on this point.

after public disclosure of the President's overall budget. Isakowitz focused on the possibility of sending humans to the Moon because he believed that it would be the most feasible option from the technical and budgetary standpoints and that it would resonate clearly with laypeople. Without hearing any specific budget estimates, O'Keefe agreed to allow Isakowitz to develop a general plan for returning to the Moon.[21]

With help from his cost estimation group, Isakowitz developed a budget estimate for lunar exploration. Using as much existing, off-the-shelf technology as possible, the team calculated that reaching the Moon in a decade with a single human flight would cost approximately $50–$70 billion. Assuming a total NASA budget of approximately $200 billion over the next 10 to 15 years, Isakowitz estimated that the Agency could carve out one quarter of the cost of a new lunar exploration initiative from existing funding.[22]

Isakowitz's estimates assumed that the Space Shuttle and International Space Station Programs would wind down both programmatically and financially. If these two major programs continued indefinitely, the Agency would have few resources available to fund existing programs aimed at future operations, such as the Orbital Space Plane (OSP), let alone provide any money for a new lunar exploration effort. That spring, after considerable tinkering on paper with NASA's projected budget, Isakowitz started to gain confidence in the feasibility of his spending plan. He ran it by the Joint Strategic Assessment Committee (JSAC), which agreed in general with the idea of developing a human and robotic exploration plan, with the Moon as the first major steppingstone and other destinations, such as libration points and Mars, as longer-term goals.[23]

On 19 May 2003, Isakowitz and Martin presented their ideas to O'Keefe in a PowerPoint presentation. Titled "New Directions: Long-Term Goals for Human Space Flight," the presentation made the case for NASA to "articulate a focused goal and develop the means for sending humans beyond low-earth orbit in the next decade." The presentation also considered the implications of the new exploration strategy for the Agency's budget and existing plans, particularly whether NASA should incorporate the new goals into its July 2003 Columbia Budget Amendment and its Integrated Space Transportation Plan (ISTP).[24]

---

21. Isakowitz interview, pp. 27–29.
22. Ibid., pp. 32–33.
23. Ibid., pp. 32–33; Martin interview, p. 15; "New Directions: Long-Term Goals for Human Space Flight" presentation, May 19, 2003, Steve Isakowitz files, p. 13, NASA HRC.
24. "New Directions" presentation, p. 2. See also Martin interview, pp. 14, 15, and especially 16 for the evolution of this New Directions presentation. On 17 December 2004, O'Keefe went back and autographed page 14 of Isakowitz's copy of this "New Directions" presentation, where he had previously checked off "Option A: Yes 'Hard' Pursuit," with a note saying "and look what

The first 14 of 19 pages of the presentation focused on the justifications for adopting a new exploration initiative. Isakowitz identified four issues that NASA's leadership needed to address to reach a decision on the matter:

- **Compelling:** Is the justification sufficient to defend the large commitment and expense?
- **Credible:** Does NASA have a solid plan and the demonstrated credentials for pursuing such an ambitious goal? Why isn't this a repeat of failed previous efforts (e.g., SEI)?
- **Urgent:** Why now?
- **Affordable:** How can the nation afford the sizable expense to undertake this goal?[25]

On the first question, loosely interpreted as whether the cause of human (in situ) exploration was compelling enough to justify the financial commitment that would be required to carry it out, Isakowitz spoke about the inherently compelling nature of space exploration, as well as the elements of his proposed exploration plan that Americans would find attractive. He quoted President Bush on the inherent human desire to explore space, and he alluded to the inspirational nature of the Apollo mission. Drawing upon earlier DPT and NEXT themes, he explained the approach NASA would take to capture the attention of the nation, piquing the interest of students and scientists seeking greater understanding of the universe. By following a steppingstones approach with humans and robotics to bring the space program to new places—which Isakowitz explained using a standard NEXT steppingstones chart—the plan would generate sustained excitement throughout the general public. Appealing to O'Keefe's broader interests as a former Secretary of the Navy and Pentagon Comptroller, Isakowitz emphasized that the benefits of the plan would extend to other national priorities, including national defense.[26]

---

happened." The Integrated Space Transportation Plan involved the Shuttle, an Orbital Space Plane, and a so-called Next Generation Launch Technology. Proposed in fall 2002, it was heralded as the "next big thing" for NASA, but the Columbia accident and related events soon overshadowed the ISTP.

25. "New Directions" presentation, p. 3.
26. "New Directions" presentation, pp. 4–5. Starting on p. 4 of this document, Isakowitz apparently jotted some insightful marginalia recording O'Keefe's reactions to specific presentation points, such as the need to highlight the national benefits of such investment, how a steppingstones approach was better than that of Apollo, the need to be sensitive to the perception of being opportunistic after the Columbia accident, and even some consideration of whether to drop a bullet point that humans and robots both bring key capabilities to space exploration.

In terms of the credibility of the plan and the credentials of NASA, Isakowitz provided more details about the steppingstones approach and highlighted the key capabilities and technologies that would be needed to carry it out. He explained the benefits of focusing on the Moon: it would be the easiest destination beyond LEO to reach and would provide opportunities for testing technologies and conducting scientific experiments on the path to more ambitious challenges. Implicitly acknowledging that NASA had lost credibility as a result of the Columbia accident, Isakowitz suggested that NASA would improve its credentials by completing "near-term confidence building milestones," such as assembling the ISS, returning the Shuttle to flight, and landing more robotic rovers on Mars. In remarking that the plan would "not be driven by manifest destiny but by scientific and technology opportunities," Isakowitz distinguished it from SEI and linked it with the core concepts of the DPT/NEXT approach. He reflected ideas from DPT/NEXT further when he mentioned that, also unlike SEI, the plan he proposed allowed for greater flexibility to deal with contingencies and funding fluctuations.[27]

O'Keefe seemed receptive to nearly all of the ideas Isakowitz discussed until they reached the section of the presentation on the question of urgency. Isakowitz argued in this section that the aftermath of the Columbia accident presented a "brief window into offering a more enduring vision beyond ISS." O'Keefe indicated that he was uncomfortable highlighting what could be seen as opportunism on the part of NASA. Isakowitz also suggested that the Agency would benefit from integrating existing programs into a larger, long-term plan rather than continuing to fund several billion-dollar programs separately, such as the Orbital Space Plane and Project Prometheus, under the assumption that they were distinct and unrelated. O'Keefe felt that such programs should be able to stand on their own, however. O'Keefe also dismissed as overstated Isakowitz's claim that the Agency needed to act quickly to make use of the knowledge gained from the Apollo program before all of the Apollo-era engineers retired or passed away.[28]

---

27. "New Directions" presentation, p. 6.

28. "New Directions" presentation, p. 9. Project Prometheus was a NASA-led, interagency effort to develop a nuclear fission reactor and high-power electric propulsion systems for robotic missions to the outer planets. The project was personally associated with O'Keefe, whose background as Secretary of the Navy and son of a nuclear submariner enabled him to understand how nuclear propulsion had changed the Navy. The project began informally in 2001 and more formally in March 2003. The project was canceled in summer 2005, after over $400 million had been spent. For more on Prometheus, see "Cautionary Tale 2: Project Prometheus" in *Launching Science: Science Opportunities Provided by NASA's Constellation System*, ed. National Research Council (Washington, DC: National Academies Press, 2008), available at *http://www.nap. edu/catalog/12554.html* (accessed 19 February 2009), p. 19; and W. Henry Lambright, "Federal

Regarding the issue of affordability, Isakowitz assumed a ballpark total of $40 billion through 2015, which was on the low end of his own Comptroller team's $50–$70 billion estimate for the first flight to the Moon. Adjusted for inflation, $40 billion in 2003 dollars equates to approximately $54 billion in 2018 dollars, much less than the $195–$243 billion in FY 2018 dollars ($20–$25 billion in then-year dollars) that NASA spent on the Apollo program. Isakowitz's figure provided for a greater balance between human spaceflight and other NASA programs than occurred during the Apollo era (a 50/50 versus 80/20 divide). Isakowitz also included a bar chart showing a "notional" budget with only minimal "offsets" to other NASA programs in later years. While accepting Isakowitz's preliminary budget estimates, O'Keefe indicated that he would not endorse any offsets this early in the planning effort. He preferred to leave it to officials in the White House to force such limitations on NASA's ambitions if they deemed it necessary.[29]

After noting that the JSAC expressed support for the concepts of an integrated human and robot exploration program that would aim to return humans

---

Agency Strategies for Incorporating the Public in Decision-Making Processes: Case Studies for NASA" (draft monograph, 18 April 2005), pp. 46–62, NASA HRC. The Orbital Space Plane effort was part of the larger Space Launch Initiative (SLI), which began in February 2001 and also included the Next Generation Launch Technology program. The OSP was intended to be a rescue and transport vehicle for ISS crews that would initially be launched atop an expendable launch vehicle and could be undocked from the ISS quickly in an emergency. After the Columbia accident in February 2003, some members of Congress pushed NASA to accelerate the OSP's development. For information on the OSP, see, for example, NASA's fact sheet on this subject from Marshall Space Flight Center, FS-2003-05-64-MSFC, at *http://www.nasa.gov/centers/marshall/news/background/facts/ospfacts.html* (accessed on 19 February 2009) and the Orbital Space Plane file 18204, NASA HRC.

29. "New Directions" presentation, pp. 11–12. On page 1271 of the *1974 NASA Authorization Hearings on H.R. 4567 Before the Subcommittee on Manned Space Flight of the Committee on Science and Astronautics*, U.S. House of Representatives, 93rd Cong., 1st sess. (Washington, DC: U.S. Government Printing Office, 1973), the total Apollo program cost is listed as $25.4 billion. The *Review of U.S. Human Spaceflight Plans Committee: Seeking a Human Spaceflight Program Worthy of a Great Nation* report (October 2009, available at *http://history.nasa.gov/AugustineCommfinal.pdf* [accessed 30 April 2018]), p. 20, puts the Apollo cost at $24.6 billion, or $129.5 billion in FY 2009 dollars. Another set of calculations, excluding categories such as tracking, construction of facilities, and potentially personnel salaries, puts the then-year total at only $19.4 billion, but $141.6 billion in 2009 dollars. For these calculations, see "apollo-budgetadjustedforinflation.xlsx" on a shared drive for information requests in the NASA HRC. This spreadsheet uses the yearly Apollo cost figures from *http://history.nasa.gov/SP-4214/app2.html* (accessed 30 April 2018) and the inflation adjustment figures previously available at *https://web.archive.org/web/20120902044105/http://cost.jsc.nasa.gov/inflation/nasa/inflateNASA.html* (accessed 13 June 2018). This inflation calculator was deposited on a shared drive for information requests in the NASA HRC (see "inflationcalculator-2012_NASA_New_Start_Inflation_Index_use_in_FY13.xlsx"). Writers often use the $20–$25 billion ballpark figure for Apollo.

to the Moon within 10 to 15 years,[30] followed by efforts to conduct operations
and set up scientific facilities at libration points before moving on to Mars,
Isakowitz wrapped up his presentation by asking O'Keefe how he wanted to
proceed. He framed three options for the Administrator: a "hard" yes, a "soft"
yes, or no. O'Keefe selected the first, most aggressive option while again warning Isakowitz that he did not want NASA to appear opportunistic so soon after
the Columbia accident. This "hard" yes entailed coupling the new human exploration plan with a budget amendment related to the Shuttle's return to flight,
seeking "seed" funding in the FY 2005 budget, looking for additional funding
in the future ("wedges" in the "outyears," in the budget argot), and adding language in an official national space policy directive authorizing human spaceflight beyond LEO.[31]

Isakowitz was pleased that O'Keefe had endorsed the ambitious, if still very
general, plan. The Comptroller felt that O'Keefe's endorsement represented a
turning point that raised the possibility that the Agency would reap a large
payoff for the previous years of DPT/NEXT planning. Several years earlier,
as an OMB budget examiner, Isakowitz had inserted the funding into the
NASA budget that supported the DPT and NEXT initiatives. Now at NASA,
Isakowitz encountered a unique opportunity to use many of the concepts developed by DPT/NEXT to pitch a plan that would represent the most important
decision for the Agency since President Kennedy announced his support for the
Apollo program in 1961.[32] Isakowitz was thrilled with the day's accomplishments and utterly unprepared for what happened next.

When he returned home from work that evening, Isakowitz excitedly and
uncharacteristically alluded to the subject of his conversation with the NASA
Administrator to his young teenage daughter. He told her that NASA was planning to send astronauts back to the Moon. Rather than the enthusiastic response
he expected, Isakowitz's daughter responded bluntly, "But haven't we already
done that?" He was stunned. He quickly realized that if the idea of returning to
the Moon did not resonate with his own daughter, his plan probably would not
gain much traction outside NASA. Isakowitz related this story to O'Keefe, who
agreed that they needed a much stronger rationale for sending humans to the
Moon. Isakowitz then worked with Gary Martin to set up an internal NASA
exercise to develop the strongest possible rationale for space exploration and to

---

30. Martin had briefed the JSAC on "Next Steps in Human and Robotic Exploration" on 13 May 2003.
31. "New Directions" presentation, p. 14.
32. Isakowitz interview, p. 44. For more of his thoughts on this presentation, see pp. 35–44 of this
     interview.

determine what the most logical next step would be, whether or not it would involve the Moon.[33]

The "Three Teams" Review

The exercise to strengthen NASA's exploration plans, which came to be known as the "Three Teams Critical Review," took place at a cloistered retreat over the course of a week, 2–6 June 2003, in Crystal City, Virginia, just 3 miles from NASA Headquarters. Gary Martin gathered NASA employees with expertise in a range of areas, many of whom had been involved with DPT and NEXT, for the Three Teams review. The Space Architect organized the NASA employees into three teams and directed each team to make the strongest possible argument for one of three alternative scenarios for space exploration.

John Mankins, who was the Manager of Advanced Concepts Studies in the Office of Space Flight and had been the primary technology person for DPT/NEXT,[34] led a team that advocated sending robotic spacecraft and then astronauts back to the Moon as preparation for human missions to Mars. Harley Thronson, an astronomer who was the technology lead in the Office of Space Science and a prominent DPT member, led a second team, which made the case for sending astronauts to assemble large telescopes and other facilities at libration points, in preparation for long human voyages to Mars, bypassing the Moon except for robotic exploration. Gordon Johnston, an engineer and spacecraft manager who was the Associate Director for Exploratory Missions in the Office of Earth Science at the time,[35] led the third team, which argued for purely robotic exploration.

In its summary presentation, Mankins's "One Frontier" team called going to the Moon a "'base camp' in the ascent to the scientific 'summit' of Mars." The team echoed DPT's call for a steppingstones approach to put humans on Mars, among other goals. Mankins's team contended that their "One Frontier" approach would be "*DRIVEN* by compelling and visionary science goals addressing profound questions that *may* be answered at Mars," including the search for life on Mars. The team also contended that lunar operations would provide opportunities for honing tools and enhancing knowledge in the fields

---

33. Isakowitz interview, pp. 45–47.
34. Mankins left NASA in 2005 after 25 years. See file 18740 (his biographical file) in the NASA HRC.
35. See DPT/VSE bioappendix files, NASA HRC, for biographical information about Johnston.

of astrobiology, medicine, geology, and Earth systems in preparation for a visit to Mars.[36]

Johnston's "Glennan" team (named after NASA's first Administrator, T. Keith Glennan) readily identified the benefits of remotely controlled (robotic) spacecraft for exploration. Remotely controlled spacecraft can be developed quickly; they are easier to launch and send to distant locations than crewed spacecraft; and considerable room exists for improving their capabilities in the future. Perhaps most importantly, they do not put astronauts' lives at risk. People may be more flexible and robust as in situ explorers, but the difficulty in human space travel is transporting astronauts between places in space that are inherently hazardous to human health. As opposed to vehicles with astronauts, robotic spacecraft can be sent on one-way space missions that are less complex and allow for greater flexibility. Despite the advantages of robotics, Johnston's team was challenged to justify an American space agency without astronauts. They believed that the fundamental difficulty of attempting to limit the space program to robotic exploration was simply that robots do not excite the general public the way that astronauts do, even though robots are much more cost-effective for accomplishing some tasks. Sending an unpiloted drone to the top of Mount Everest, for example, might be impressive technically, but it would not be as engaging and dramatic as people climbing to the summit of Earth's highest peak.[37] Gaining funding would be difficult, which the team acknowledged with an aphorism well known in the space community, "no Buck Rogers, no bucks."[38]

Thronson's "Search for Life" team contended that a "direct to Mars" approach provided the strongest scientific case of the three options because it would allow NASA to reach the most scientifically and publicly compelling destination soonest. The team argued that NASA could use the Moon for scientific exploration without any direct astronaut involvement. The Moon had limited public appeal, in any case, because it did not represent a new challenge

---

36. "One Frontier—One NASA" presentation ("OneFrontier-Moon.ppt"), in Three Teams electronic materials from Harley Thronson, 6 June 2003, NASA HRC. The base camp quotation is from p. 13, and this term is also used on p. 1. The "driven" quotation, with emphasis in the original, is from p. 6. For common tools, see p. 32. For lunar and Martian science commonalities, see pp. 37–39.

37. "Glennan Team," Three Team Critical Review, 6 June 2003, Three Teams electronic materials, Harley Thronson files, NASA HRC. The references are from pages 10, 16, 11, 6, and 16 again, respectively.

38. While NASA spends billions of dollars each year on both human and robotic spaceflight, the argument behind this aphorism is that without the public allure of human spaceflight, there would be little public or congressional support for NASA's robotic scientific efforts and perhaps even for NASA to exist at all.

or destination. While some members of this team expressed interest in having astronauts assist with the building of large scientific observatories at libration points, others believed that such missions could be accomplished with limited human participation.[39] Moreover, Thronson noted that long human voyages to Mars would require extensive preparation possible only in free space, not on the lunar surface. The team's larger point was that NASA already had enough experience to begin planning to send humans to Mars. Thronson, according to his own recollections, paraphrased a famous quotation attributed to Napoleon Bonaparte about maintaining focus by contending that "if NASA intends to go to Mars, then *go to Mars!*"[40] In other words, the Moon was a significant distraction, if not a total dead end. If Mars was the real objective, then NASA should stop sneaking up on it and begin moving toward it with speed and purpose.

After the week of vigorous debate, the participants voted informally for the most persuasive option and presented their results to Martin and Isakowitz. The straw poll resulted in a tie between the option of going to the Moon followed by a trip to Mars and the option of bypassing the Moon and pressing on directly to Mars. Interestingly, the three teams' members agreed that the third option of sending only robots beyond LEO "made the strongest logical case" in terms of science, but they concluded that science alone was not the only criterion to consider in determining whether to send humans beyond LEO.[41] The Three Teams participants felt that the DPT science-driven approach was a "necessary but not sufficient" organizing principle for exploration; they advocated for incorporating more intangible concepts, such as "discovery" and "adventure." The three teams also agreed on the importance of using a steppingstones approach, establishing technological capabilities first, and leaving the selection of destinations to a later date.[42] The main lesson that the participants learned from the exercise

---

39. "The Search for Life" team presentation ("Search_for_Life-2.ppt"), Three Teams electronic materials, Harley Thronson files, NASA HRC. Regarding observatories at libration points, on one hand, one of the "ground rules and assumptions" for Thronson's team was that "[s]cience facilities at the libration points will achieve very high priority science for NASA" (p. 3). Yet one of their conclusions was that "[l]ibration point missions can be achieved either autonomously (limited human intervention) or are of lower scientific value" (p. 9).

40. Napoleon Bonaparte said "If you start to take Vienna—take Vienna." See, for example, *http://thinkexist.com/quotation/if_you_start_to_take_vienna-take/195324.html* (accessed 4 March 2009). Thronson recalls verbally paraphrasing this quotation in a summary presentation, but there is no apparent written record of this quotation.

41. "Three-Team Critical Review: Team Reports, Emerging Findings, and Initial Recommendations," presented to Space Architect, 6 June 2003, hard copy in Three Teams files from Harley Thronson, pp. 18 and 21, NASA HRC.

42. "Three-Team Critical Review," p. 15; DPT/NEXT "'Lessons Learned' History" PowerPoint file, Three Teams files, p. 13.

was that no simple justification existed for a broad exploration program featuring humans in space along with robotic spacecraft.[43]

For Gary Martin, the Three Teams exercise was one of many strategic planning studies that he participated in, commissioned, or reviewed. He felt that it was useful in clarifying roles for science and for human spaceflight, especially regarding the Moon. The exercise helped to convince Martin that the Moon was a logical first step for exploration because NASA could not afford to send humans directly to Mars and the public would never understand the logic of going to libration points, even as a preparation for human missions to Mars. Scientists could easily conceive of interesting experiments to conduct on the lunar surface if NASA chose to return to the Moon, yet virtually all lunar science could be accomplished with robots, as Martin recalled Ed Weiler (among others) warning.[44]

The Three Teams exercise was relatively unique for NASA in that it was an organized, deliberative face-to-face debate over competing options for the Agency's future. Nonetheless, the Crystal City exercise did not and probably could not resolve the perpetual debates about the roles of robotic and human spaceflight. Having the participants openly debate the merits and drawbacks of each approach before one became a formal Agency policy helped key officials understand the strengths and weaknesses of competing ideas and proposals.

The White House Rump Group

While NASA continued its internal debate on future options, at a White House meeting on 26 June 2003, Andrew Card, the President's Chief of Staff, asked for NASA, the White House, and other relevant agencies to form an informal working group to consider options for redefining NASA's mission. With the formation of what was known as the "Rump Group," the informal teams at NASA and the White House that had been working in isolation on proposals for future space exploration began collaborating for the first time. Meeting approximately every week, the Rump Group provided a forum for NASA and White House staff members to vet ideas and to work toward the common goal of identifying the most salient approach to space exploration in the future.[45]

NASA placed three representatives on the committee: Isakowitz; Mary Kicza, the Associate Administrator for Biological and Physical Research;

43. DPT/NEXT "'Lessons Learned' History" PowerPoint file, p. 13.
44. Martin interview, 16 February 2006, pp. 202–222.
45. John Schumacher handwritten notes, 26 June 2003, pp. 1–2, Rump Group folder, Schumacher files, NASA HRC.

and John Schumacher, O'Keefe's Chief of Staff. O'Keefe presumably picked Isakowitz for his budget skills, political acumen, discretion, and OMB experience. Schumacher brought substantial political skills to the process, as well as experience coordinating interagency and international negotiations. Kicza, an engineer who led a scientific organization, had a strong technical background, although she was not a close confidante of O'Keefe. Lisa Guerra, who was then Kicza's special assistant for strategic planning, provided Kicza with background information (although Guerra did not know why Kicza wanted it or what would be done with it). O'Keefe provided Isakowitz, Kicza, and Schumacher with few details about what he wanted from the process other than for them to devise a new space exploration plan of major significance.[46] Besides reporting back to O'Keefe, the three NASA representatives to the Rump Group also debriefed Fred Gregory and Gary Martin.[47]

Participants from the White House included Gil Klinger from the NSC and Dave Radzanowski from OMB. Jim Marrs, from the Vice President's staff, attended several early meetings.[48] Bill Jeffrey of OSTP served as the de facto leader of the Rump Group. Brett Alexander, who had strong disagreements with Jeffrey and other OSTP managers about the direction of the civilian space program, was conspicuously absent from the group. He took a vacation around 4 July and was not included in the meetings upon his return. Although he spoke with Jeffrey and Klinger and saw the written products they were working on, he did not attend Rump Group meetings for about six weeks in midsummer; he then resumed his participation in the administration's deliberations in late August. John Marburger, the head of OSTP, considered Jeffrey the "most senior person who was highly knowledgeable about NASA" among the group of individuals involved in the White Office effort. Marburger also held Alexander and Klinger in high esteem and thought the two were well qualified to lead teams at the working group level.[49]

Before the Columbia accident, President Bush was preoccupied with fighting terrorism and gave no indication that he had spent much time thinking

---

46. O'Keefe interview, pp. 22–24; Schumacher interview, pp. 13–16, NASA HRC.

47. Martin interview, p. 63.

48. Because the Rump Group was informal—meaning the group lacked a formal charter and no high-level recognition attended the initiation of the working group or completion of its activities—no complete list of members appears to exist. See Schumacher's handwritten notes, Rump Group folder, Schumacher files, NASA HRC, for an informal list of members.

49. Alexander interview, pp. 19–20; John Marburger, interview by Asner and Garber, NASA Headquarters, Washington, DC, 22 January 2010, p. 9, NASA HRC. On p. 10 of this interview, Marburger also notes that at that time, he and his Associate Director for Technology, Richard Russell, kept in touch with Jeffrey and Alexander at least every week on civilian space policy.

about civilian space policy. Marburger viewed the President as noncommittal about space and seemingly of the perspective that human spaceflight was too expensive. After the accident, Marburger received no private guidance from the President, leaving only what Bush had expressed publicly in his support for human spaceflight as a basis for approaching policy discussions. Bush was content to let the policy process run its course, entering it only at the end to make a final decision. In a later interview, Marburger identified tensions between the White House and NASA as rooted in their different roles in the policy-making process. With the White House fully in command and staffers at the NSC guiding the policy-making process, NASA was aware that the end result might be a policy the Agency did not want. Given NASA's lack of control over the process, NASA leaders were wary of putting forth or endorsing any proposal that was not fully worked out.[50] Marburger outlined several general aims for the Rump Group in a 30 June email to O'Keefe and Josh Bolten, the White House Chief of Staff who succeeded Andrew Card. He directed the group to prepare a draft paper outlining a "compelling, long-term vision" for space exploration for the White House and NASA to review within a month. Marburger expected the group to develop several possible space exploration options, ranging "from conservative to aggressive," and to articulate a rationale for continuing to fly humans in space. O'Keefe agreed with Marburger's major points and recommended that the group attempt to build on earlier interagency efforts, prior to the Columbia accident, aimed at developing a new National Security Policy Directive.[51]

The Rump Group reached agreement in its early meetings on a few core assumptions that would inform any new long-term exploration strategy. First, NASA would safely return the Shuttle to flight and complete the ISS assembly to support a crew of six. Second, NASA would retire the Shuttle after completion of the ISS. Last, the administration should avoid raiding other NASA programs to pay for additional human spaceflight expenses.[52]

50. Marburger interview, pp. 7, 16, 18, 23. According to Marburger, there was "tension" between the White House and NASA because "NASA had this inclination—a natural inclination that you find in Washington—not to talk too much about thoughts that hadn't been completely formulated." Marburger felt this was unfortunate because it reduced the opportunities for having honest and creative discussions about the direction of the Agency. (See p. 16 of his interview.)

51. O'Keefe to Marburger and Bolten, 30 June 2003, Vision Development emails folder 1 of 3, box 3, John Schumacher files, NASA HRC; Marburger to O'Keefe and Bolten, 30 June 2003, Vision Development emails folder.

52. "Decision Memorandum: Future of Human Space Flight," 9 July 2003 draft outline, p. 1, Rump Group folder, Schumacher files, NASA HRC.

The group considered several options for structuring their hypothetical exploration program. The first option called for terminating all human spaceflight efforts after completing the Shuttle Program and either reducing NASA's overall budget by an equal amount or using the savings for robotic programs. The second option envisioned NASA developing separate new launch vehicles for crew and cargo if funding remained after the termination of the Shuttle Program and completion of the ISS assembly. The third option involved pushing human spaceflight beyond low-Earth orbit—with the caveat that the ambitious exploration program would differ from SEI and that a range of options would exist for the funding and pacing of the program.[53]

The Rump Group leaned toward the third option. Team members believed that the White House should reiterate its support for returning the Shuttle to flight and completing the ISS assembly soon after the release of the CAIB Report. At a suitable high-level event at a later date, the President could announce a new long-term plan for human exploration beyond LEO. The Rump Group members likely would tie any decision regarding whether to adopt a more conservative, lower-cost approach or a more aggressive, higher-cost option to the established annual budget process.[54]

As for the rationales for human spaceflight that Marburger requested, the Rump Group presented lists of possibilities, ranging from material to metaphysical benefits, that would have resonated with anyone familiar with the literature on the justifications for human space exploration. On the material side, the Rump Group focused on the possibilities of maintaining and improving economic and national security through technological progress, technological spinoffs, and the expansion of science and engineering education for future generations. Less tangibly, the group mentioned the inspiration future generations would receive from investments in space today and the importance of satisfying the supposedly deep human yearning for exploration. Science and diplomacy also entered the discussion, with the themes of international cooperation, national prestige, and the possibility of human space exploration opening up vast possibilities for scientific research emerging as important rationales as well. Most of these rationales, however, were less than robust, the Rump Group concluded. Many of the rationales did not necessarily require human spaceflight, and others would be hard to quantify and/or justify. The credibility of the rationales, the group concluded, rested on the inspirational qualities of human spaceflight, the

---

53. "Decision Memorandum: Future of Human Space Flight," p. 2.
54. Ibid., p. 2, and undated, unpaginated attached sheet, "Rationale for Human Spaceflight," Schumacher files, Rump Group folder, NASA HRC.

relative value of spinoffs compared to direct investments, and whether humans held any advantages over robots for identifying alien life in space.[55]

To put their task in perspective, the Rump Group members also looked at the consequences of abandoning human spaceflight. Addressing fears about competition from the Russians and Chinese, the Rump Group concluded that "we don't actually have to be first in everything—let them throw their money down a hole." The Rump Group members agreed that it was not their responsibility to worry about the impact of canceling human spaceflight and that the decision should be considered only at the presidential level. Regarding U.S. commitments to international partners in space (most notably on the ISS), the Group members concurred that it would be reasonable and "not unprecedented to terminate a relationship that leads you to doing something stupid. Some or all of the resources freed up could go to other international efforts."[56] Reflecting a post–Cold War perspective, the Rump Group was not initially inclined to view human spaceflight as a critical component of either national security or international relations.

In the final white paper designed to help the White House principals reach a decision on the future of space exploration, the Rump Group followed up on the themes mentioned above and raised several key questions. First, what would be the final ISS configuration? Second, what kind of launch vehicle system would replace the Shuttle and when? Third, should the United States maintain an independent and continuous human spaceflight capability? Finally, should the nation pursue human spaceflight beyond LEO, and if so, how aggressively? The Rump Group produced this decision paper in July 2003, anticipating the release of the CAIB Report the following month and the start of a larger and more publicly visible policy debate.[57]

The decision paper provided background points to set the context for its various exploration proposals. The paper first explained that NASA and the human spaceflight program were in a tenuous position. With only three Shuttle orbiters remaining, the loss of another orbiter would cripple both the Shuttle and ISS Programs.[58] Following the conventional wisdom of the space community, the decision paper then mentioned that NASA could not afford to fund any major

55. "Products from 'Rump' Group" document, undated but attached to "Decision Memorandum: Future of Human Space Flight," p. 2.

56. Ibid., p. 2; "Decision Memorandum: Future of Human Space Flight," p. 2, and undated, unpaginated attached sheet "Rationale for Human Spaceflight."

57. "Decision Paper: U.S. Human Space Flight Vision and Issues," July 2003, Isakowitz folder, pp. 1–7, NASA HRC.

58. After the Challenger accident in 1986, Congress readily appropriated more than a billion dollars for NASA to build a replacement orbiter. Almost 20 years later, this was unlikely to hap-

new human spaceflight endeavors with its existing budget unless the Shuttle and/or ISS Programs were scaled back significantly or terminated early. Third, the Rump Group attempted to gain agreement on the importance of developing a new vision by reminding readers of the report that NASA had no long-term plans for human spaceflight beyond the ISS.[59]

Marburger expected the group to focus its efforts on developing a range of options for exploration. In a summary chart, the decision paper presented seven alternative exploration plans, ranging from most aggressive (abandoning current infrastructure and starting immediately on a new program) to implementing a new plan with a somewhat undefined "Advanced Technology Vehicle" after a "brief hiatus."[60]

Marburger urged NASA and O'Keefe to focus immediately on future missions, rather than taking a passive, status quo approach that would build on existing ISS commitments. The Rump Group came up with three basic options: end human spaceflight and retire the Shuttle; focus on the ISS; or expand human spaceflight beyond LEO without repeating the perceived mistakes of SEI. Not surprisingly, they dismissed the first two plans as "defensive" and instead recommended the third plan as a way to seize the policy initiative.[61]

The CAIB Report

The Columbia Accident Investigation Board issued its eagerly awaited final report on 26 August 2003.[62] The report was remarkable in several respects. It was well written and explained technical issues clearly for readers unfamiliar with such technical issues and situated these issues in their larger social, organizational, cultural, political, and historical contexts. It declared that the root causes for the Columbia accident were both technical (foam struck the orbiter and damaged the thermal protection tiles) and cultural (NASA as an organization had forgotten the lessons of the Challenger accident, and its managers were not fostering open enough internal communication, especially in allowing, let alone encouraging, the airing of dissenting opinions by working-level engineers and technicians). Some critics simply believed that NASA never "learned the lessons of Challenger" and merely paid them lip service.

---

pen again, given that the Shuttle Program was generally thought to be a mature operational program until the Columbia accident happened.

59. "Decision Paper: U.S. Human Space Flight Vision and Issues," pp. 1–7.

60. Ibid., pp. 1–7.

61. Marburger to O'Keefe and Bolten, 24 July 2003, Vision Development emails and Rump Group folder, box 2, John Schumacher files, NASA HRC.

62. The complete report is available at *http://history.nasa.gov/columbia/CAIB_reportindex.html*.

Chapter 9 of volume I, titled "Implications for the Future of Human Space Flight," departed from the arc of the typical technical accident report. While some individuals, such as Klinger and Alexander, were concerned initially about the CAIB wading into the public arena and prescribing policy, such concerns dissipated once they read chapter 9.[63] Chapter 9 made two fundamental points about planning for human spaceflight. First, since the completion of the Apollo program, the rationale for human spaceflight was not sufficient to justify large and sustained budgets. Whereas the Cold War Apollo program was linked to national security priorities, human spaceflight following Apollo had no clear relationship to any significant national priority, military or otherwise. As the CAIB Report explained:

> Since the 1970s, NASA has not been charged with carrying out a similar high priority mission that would justify the expenditure of resources on a scale equivalent to those allocated for Project Apollo. The result is that the agency has found it necessary to gain the support of diverse constituencies.... NASA has usually failed to receive budgetary support consistent with its ambitions. The result...is an organization straining to do too much with too little.[64]

Without a connection to an important national priority, particularly to national security concerns, the space program had no special claim on national resources and was subject to the whims of the normal political process and competition for resources. NASA rarely received budget increases in recognition of the importance of its core programs.

The second basic point was that the government had failed to make a serious and sustained commitment "over the past decade to improving U.S. access to space by developing a second-generation space transportation system," despite much clamoring in the space policy community for a new, more reliable launch vehicle. Precisely because NASA's human spaceflight program lacked a compelling mission, both Congress and the White House had been reluctant to commit the additional billions of dollars required to develop a new launch vehicle. The space community attempted to respond to this absence over the decades by developing revolutionary new "leapfrog" technologies, but developers failed repeatedly to overcome technical barriers. The report declared, *"previous attempts to develop a replacement vehicle for the aging Shuttle represent a failure of national leadership"* [italics in original]. Given the lack of a major breakthrough, the CAIB recommended that "the country should plan for future

---

63. Klinger interview, p. 14; Alexander interview, pp. 34–35.
64. CAIB Report, vol. I, p. 209.

space transportation capabilities without making them dependent on technological breakthroughs." In other words, the time had come for NASA and the space community to stop investing in the development of vehicles that promised performance beyond existing capabilities and to instead focus on developing new vehicles based on proven and reliable designs.[65]

Given the age and condition of the Shuttle fleet, as well as flaws with the basic design of the Shuttle, the report argued that "it is in the nation's interest to replace the Shuttle as soon as possible as the primary means for transporting humans to and from Earth orbit." The CAIB agreed with NASA that the Orbital Space Plane was only an interim solution for placing astronauts in space. The government would need to provide a sustained national commitment, without predetermined funding limitations, to develop a truly robust and viable vehicle.[66]

Alexander and Klinger found the CAIB's assessment of the root causes of the Columbia accident convincing. They also felt that the CAIB was justified in calling for national leadership to provide a consensus in support of whatever path the human spaceflight program would take. As Alexander noted, the lack of leadership extended to "both ends of Pennsylvania Avenue, and it's both [political] parties, and it's 30 years of blame. Many administrations, many Congresses. It really is this lack of vision." These administration members saw the CAIB Report as a confirmation of their shared view that the previous 30 years of spaceflight represented a costly and unnecessary diversion from a bolder path of exploration.[67]

Political leaders in the Bush White House, according to Alexander, believed that the accident signaled a need for a fundamental change in the human spaceflight program. Alexander suspected that senior members of the White House believed that history would judge them poorly if they failed to address the CAIB's two key points in chapter 9: establishing a new human exploration program with a compelling mandate linked to national priorities and ensuring a sustained financial commitment to developing a new launch vehicle. Alexander sensed concern about the possibility of another accident occurring during President Bush's second term if the White House failed to support a change in NASA's safety culture and to deal with the proximate technical cause of the Columbia accident.[68]

65. Ibid., p. 209. See also Logsdon, "'A Failure of National Leadership': Why No Replacement for the Shuttle?"
66. The quotation is from the CAIB Report, vol. I, pp. 210–211.
67. Alexander interview, p. 34.
68. Ibid., p. 35.

In this sense, the CAIB inspired further debate about Shuttle safety and raised expectations for a shift in national space policy. While NASA, OMB, OSTP, and others at the White House were considering a new mission for NASA for several months, the planning efforts remained subdued until the release of the CAIB Report. The report provided support for a new space policy and legitimized the interagency planning effort already under way. It affirmed the importance of moving forward with human spaceflight and allayed the concerns of O'Keefe and others that NASA might appear opportunistic if it pushed for an expansion of the Agency's mission after the accident.

6

"BOLD IN VISION AND CHEAP IN EXPENSE"[1]

BY EARLY AUGUST 2003, the leaders of the interagency discussions reached the conclusion that the Rump Group was not likely to produce a major policy decision.[2] At a 12 August meeting with Marburger and Stephen Hadley, the Deputy Assistant to the President for National Security Affairs, White House Chief of Staff Andrew Card gave responsibility for overseeing the development of the Vision to Hadley and the National Security Council.[3] To the extent that the National Security Council was the lead player in policy development in the executive branch, the decision to give the NSC control over the process mainly signaled the White House's serious intent to develop a new exploration strategy for NASA. The NSC was unique among executive branch organizations in that it provided a clearly defined process for bringing major policy initiatives to presidential approval. Council staff also held responsibility for space-related

---

1. Isakowitz interview, p. 73. The full quotation from Isakowitz was, "[T]he White House was pretty consumed with not wanting to spend much money. Ultimately, the White House loved a vision that was bold in vision and cheap in expense."

2. The Rump Group meetings laid out multiple scenarios and raised many important questions, but the participants did not reach agreement on which options or exploration scenarios were most viable. In this respect, OSTP's failure to provide a clear path for reaching a decision convinced those interested in seeing a new policy produced that another organization needed to assume leadership of the process. Alexander interview, pp. 21–23; Klinger interview, p. 21.

3. Isakowitz interview, pp. 60–61; Schumacher interview, p. 22; Klinger interview, p. 21. According to Isakowitz and Schumacher, Sean O'Keefe approached Hadley about bringing the vision development into the NSC decision-making structure in order to ensure that senior officials had a venue for considering the recommendations that emerged from staff-level discussions. O'Keefe also had greater confidence that NASA's ideas would get a fair hearing with his old DOD colleagues, Miller and Hadley, in charge.

Dr. John Marburger was President George W. Bush's Science Advisor. (Brookhaven National Laboratory)

issues that involved multiple government agencies, at least since the beginning of the Bush administration.[4]

Although he was consumed with larger national security issues, particularly wars in Iraq and Afghanistan, Hadley accepted Card's request without protest. Hadley's subordinates, who would do the bulk of the work, were less enthusiastic about the opportunity. In Klinger's words, "Every chance I got I tried to push this away because I was worried it would be a train wreck."[5] While Klinger's reaction may appear odd, given how instrumental he had been earlier, both he and Miller felt that the NSC was the wrong organization to lead the process. They pointed out that the Security Council was not in the civilian space business, and Klinger readily admitted that he had no expertise in the area. More importantly, Klinger and Miller already had too many responsibilities. They were preoccupied with developing a new set of National Security Presidential Directives (NSPDs) on space.[6] Upon hearing their protests, however, Hadley apparently asked, "What is the right thing to do for the President?" Klinger and Miller could not refuse Hadley and thus reluctantly agreed to take charge of the policy development process.[7]

---

4. Alexander interview, pp. 21–23, 26; Klinger interview, pp. 21–23; Isakowitz interview, p. 61. On the National Security Council policy-making structure of the Bush administration, see National Security Presidential Directive 1 (NSPD-1), 13 February 2001, available at *https:// www.georgewbushlibrary.smu.edu/~/media/GWBL/Files/Digitized%20Content/2014-0390-F/ t030-021-012-nspd1-2-20140390f.ashx* (accessed 23 May 2017).

5. Klinger interview, p. 21.

6. Alexander, on the other hand, was glad that Klinger was offered the opportunity to lead the process at the staff level and urged him to accept the work, going so far as to argue that refusing to accept responsibility for leading the development of the vision would kill both the vision and their joint efforts to gain acceptance for the remaining NSPDs (Alexander interview, p. 22). The Bush administration issued at least six NSPDs on space. In addition to NSPD-15, which initiated a National Space Policy Review, and the vision directive, NSPD-31, "U.S. Space Exploration Policy," President Bush signed NSPDs titled "Commercial Remote Sensing Space Policy" (NSPD-27, on 25 August 2003); "Space-Based Position, Navigation, and Timing Policy" (NSPD-39, on 8 December 2004); "Space Transportation Policy" (NSPD-40, on 21 December 2004); and "National Space Policy" (NSPD-49, on 31 August 2006). See *http://www.fas.org/irp/offdocs/nspd/space.html* (accessed 14 October 2011) for a list of NSPDs.

7. Alexander interview, p. 23.

Deputy National Security Advisor Stephen Hadley and White House Domestic Policy Advisor Margaret Spellings receive awards from NASA Administrator O'Keefe for their work crafting the VSE. (NASA)

Soon thereafter, the Domestic Policy Council (DPC) was brought in to serve in a co-leadership capacity, although the NSC remained dominant. Klinger later claimed that the White House split the leadership between the NSC and DPC to allay concerns that might arise regarding the propriety of allowing a national security organization to set policy for a civilian agency. At the highest levels, then, responsibility for developing the vision rested with Stephen Hadley and Margaret Spellings, the Assistant to the President for Domestic Policy.[8]

The Policy Process Begins in Earnest

Under this new arrangement, the process proceeded over the next several months along two tracks: working group meetings at the staff level and Deputies Committee meetings with all of the relevant senior players, usually in the White House Situation Room.[9] The cochairs of the working group, Frank Miller of the

8. Klinger interview, p. 22; NSPD-31, 14 January 2004.
9. The Deputies Committee vetted and reviewed national security policy issues in preparation for further discussion at Principals Committee meetings and final decision by the President at National Security Council meetings. Regular members included representatives in

President George W. Bush's Chief of Staff, Andy Card. (White House photo by Tina Hager, *https:// georgewbush-whitehouse.archives.gov/government/ card/04.html*)

NSC and Diana Schacht of the DPC, delegated responsibility for coordinating the effort to Gil Klinger. With much of the team drawn from the Rump Group, the NSC/DPC working group was familiar with the questions facing NASA and the Bush administration. Not surprisingly, others involved in the meetings, including representatives from DOD and the National Reconnaissance Office (NRO), appear to have participated primarily to protect their own interests—to ensure that any policy the Committee might put before the President would not limit the scope of their organization's responsibilities or negatively impact their major programs.[10]

While the working group members typically were familiar with space issues, some of the Deputies had not been involved in space policy prior to the first Deputies Committee meeting. Deputies from the DPC and the Council of Economic Advisors, according to Alexander, "didn't have any space knowledge." Given the range of issues they typically consider, to be fair, it is not uncommon for officials at the highest levels of the policy-making process to

---

Deputy or Under Secretary positions at the Departments of State, Treasury, and Defense; the Deputy Directors of Central Intelligence and the Office of Management of Budget; the Deputy Attorney General; the Vice Chairman of the Joint Chiefs of Staff; the National Security Adviser; and a handful of other high-ranking officials in the White House. The membership of the Committee would expand to include other senior government officials as needed based on the topic under consideration, such as when the NASA Administrator or Deputy Administrator would be included in discussions regarding space policy. See NSPD-1 from 13 February 2001, available at *https://www.hsdl.org/?abstract&did=462808* (accessed 26 June 2018).

10.   Alexander interview, pp. 24–25. Other members of the working group included Alison Boyd from the DPC, Jim Marrs from the Office of the Vice President, Bill Jeffrey and Brett Alexander from OSTP, David Radzanowski from OMB, Eric Helland and Harvey Rosen from the Council of Economic Advisors, John Schumacher and Steve Isakowitz from NASA, Bud Rock and Ken Hodgkins from the Department of State, Robert Dickman and Robert Kehler from the NRO, Thomas Scheber and Dave Trottier from DOD, David Smith from the Joint Chiefs of Staff, and Randy Soderholm from the Office of the Assistant Director of Central Intelligence (list of VSE Executive Office of the President [EOP] working group members, Radzanowski files, NASA HRC).

Gil Klinger receives a NASA award for his work on the Vision for Space Exploration policy. (NASA)

Brett Alexander was a key architect of the VSE from his position at the White House Office of Science and Technology Policy. (NASA)

have little expertise, or not even a basic level of knowledge, about many of the issues that come before them. While Marburger was somewhat disappointed that many of the people considering the new space policy were not technically inclined, he also believed that the issues in question were too complex to be boiled down to yes-or-no decisions. Whether because of the Deputies' lack of space knowledge or the complexity of the issues involved, the officials devoted a significant amount of time to discussion and deliberations.[11]

To frame the policy debate, Harriet Miers, then the White House Deputy Chief of Staff for Policy, worked with Alexander and Klinger (and possibly Frank Miller, the NSC Senior Director for Defense Policy and Arms Control) in late August 2003 to prepare a set of background questions for a future Deputies Committee meeting.[12] On Tuesday, 19 August, Alexander forwarded the list of questions to Isakowitz, Schumacher (whom O'Keefe had appointed Chief of Staff a month earlier), and representatives of the Department of State, Central Intelligence Agency (CIA), DOD, and Joint Chiefs of Staff. He requested

---

11. Alexander interview, p. 24; Marburger interview, pp. 20–21, 51. In his interview, Alexander indicated that the Deputies met between 5 and 10 times. He noted that having one such Deputies meeting to reach a decision is a "big deal," having two such meetings is reserved for "very serious" issues, and having 10 would be "unheard of." Records examined for this study indicate that the Deputies held formal meetings at least five times.

12. Alexander interview, p. 26.

responses within 10 days, by the close of
business on Friday, 29 August. The 15
questions fell into five main categories:

- Genuine NASA requirements for
  human spaceflight
- The benefits of human spaceflight
  and human exploration beyond LEO
- International obligations as a result
  of the ISS
- Requirements and alternatives for
  completing the Station
- The capabilities of the Shuttle and
  the Orbital Space Plane

NASA Chief of Staff John Schumacher
played a significant role in interagency
deliberations on the VSE. (NASA
Schumacher_20030808)

Alexander tasked NASA with answering all of the questions, the military and
intelligence agencies only with explaining
their own requirements for human spaceflight, and the Department of State with
four questions related to the nation's international obligations.[13]

Schumacher immediately sent the questions down his chain of command for
assistance on NASA's response. Gary Martin distributed different questions to
the relevant individuals within the Agency, some of whom had worked on DPT.
Schumacher gave responsibility for preparing the overall package to the organization he had led for the past seven years, the Office of External Relations
(OER), presumably because he had confidence in OER's abilities and because
several questions involved international topics. Schumacher notified O'Keefe
and NASA Associate Administrator Fred Gregory by email late in the after-
noon on 19 August about the request. Gregory immediately responded with
encouraging words; O'Keefe, who was then on vacation in upstate New York,
responded a few minutes later that Retha Whewell, his executive assistant, had
informed him about the request the prior day. O'Keefe considered the questions
biased against NASA and human spaceflight, which he attributed to the fact
that, as he understood it, Brett Alexander had developed them.[14]

13. Brett Alexander to John Turner et al., "Human Space Flight Questions," 19 August 2003, Joe
    Wood files, NASA HRC.
14. Gregory to Schumacher, 21 August 2003; O'Keefe to Schumacher, 21 August 2003, both in
    Vision Development emails folder, John Schumacher files, NASA HRC.

Whether O'Keefe's perception of Alexander as biased stemmed from personal experience with him or was merely a product of O'Keefe's overall judgment of OSTP, the questions could be read as constructed in such a way as to force NASA to lay out options that would limit the scope of a new vision. Questions about the purpose, utility, and expected social benefits of human spaceflight gave NASA an opportunity to wax eloquently about the value of human exploration, but they also evoked thoughts about costs and the expected payoffs from exploration in comparison with other government programs. Questions about NASA's legal, contractual, and diplomatic commitments to complete the Space Station, alternatives for meeting those obligations, and the value of the ISS and the Shuttle for exploration beyond LEO raised the issue of terminating Shuttle operations and ending the ISS sooner than the Agency desired.[15] Whether Alexander sought to lead the agencies in any direction, the questions he posed hinted at some of the conflicts—over cost, the termination of the Shuttle, and NASA's commitment to international partners—that would emerge in the coming months.

At a formal meeting on Tuesday, 2 September, at 6:00 p.m. in the White House Situation Room, the Deputies discussed the broad motivations and goals of the vision development process. Attendees at the meeting included Hadley, Marburger, O'Keefe, Deputy Secretary of State Richard Armitage, Joel Kaplan from OMB, Harriet Miers, John Schumacher, Frank Miller, and Gil Klinger. The Deputies agreed to a series of broad statements regarding the process, including the following:

- The White House is in the process of creating a broad new vision for space exploration.
- The Columbia accident focused the administration on reconsidering NASA's mission.
- The final report of the CAIB would inform the policy development process.
- The U.S. space program is important to all of humanity.
- International partners are important to ongoing U.S. space efforts.
- Overall administration policy objectives would set the tone for space policy.[16]

Whether these were talking points for public consumption or points of agreement to set the tone of future meetings, they generated no major disagreements among the Deputies.

---

15. Brett Alexander to John Turner et al., "Human Space Flight Questions."
16. O'Keefe calendar, 2 September 2003, Sean O'Keefe files in Vision files, NASA HRC; Schumacher to O'Keefe and Paul Pastorek, 3 September 2003, in Vision Development emails folder 1 of 3.

The tone of the meeting was cordial, but Gil Klinger sensed trouble. O'Keefe wanted a budget increase to develop the new plan on a foundation of existing programs—Shuttle, Project Prometheus, and the Orbital Space Plane. OSTP, on the other hand, was pushing for a program that did not venture far from existing budget projections. While the differences between Marburger's and O'Keefe's positions may have been muted due to the nature of the meeting, Klinger thought that the two "were on completely different planets" and subsequently attempted to explain the major differences between their views, seemingly to no avail, to Miers, Spellings, and Hadley in Harriet Miers's office.[17]

The 2 September meeting, nonetheless, opened the door for more systematic consideration of the alternatives available for redefining NASA's mission. At the behest of some of the agencies involved in the process, the Deputies gave all of the agencies a chance to develop their own proposals. Over the next three weeks, teams at NASA, OMB, OSTP, and CEA developed brief white papers and PowerPoint presentations following a standardized format in preparation for a 25 September Deputies Committee meeting.[18]

Discussions at the working group level provided an opportunity for the teams to circulate their proposals and to get a sense of the major differences among them. The OMB proposal, "Mars in Our Lifetime," raised serious concerns at NASA. It called for immediately terminating the ISS and Shuttle Programs in order to save money to begin development of heavy-lift and crew transport vehicles for Mars exploration, with the Moon as a first steppingstone. From her position as head of biomedical research at NASA, Mary Kicza contended that since the radiation exposure of a long trip to Mars would be a major concern for astronauts, the ISS would actually be a much better laboratory or steppingstone than the lunar surface. She echoed one of the concerns expressed months earlier, during NASA's Three Teams competition, when she asserted that preparing for a mission to Mars without the benefit of microgravity research on the ISS would increase the risk that astronauts would not be able to function properly on Mars after a six- or seven-month journey and that their ability to lead productive lives after returning to Earth would diminish.[19] Thomas Williams, a White House Fellow in O'Keefe's office, also weighed in, arguing that allowing a long gap between the termination of Shuttle operations and the resumption of human spaceflight with a new vehicle several years later would result in a loss of

17. Klinger interview, pp. 18–25, quotation on p. 18.

18. Isakowitz interview, p. 69; Alexander interview, p. 27.

19. Mary Kicza to John Schumacher, 16 September 2003, Vision Development emails folder 1 of 3, box 3; David Radzanowski to Frank Miller et al., 15 September 2003, attachment, "Vision Matrix final2.doc," Vision Development emails folder 1 of 3, box 3.

technical capabilities among members of the astronaut corps.[20]

A week before the 25 September Deputies Committee meeting at the White House, O'Keefe and Schumacher appeared confident that the Deputies ultimately would accept the NASA proposal over the alternatives. Key departments and offices that had not developed alternative proposals, including the Department of State, DOD, and the Office of the Vice President, expressed support for NASA's plan in discussions with Schumacher. Representatives from the Department of State promised to emphasize the importance of maintaining international alliances and meeting commitments to foreign partners on the Space Station, while Schumacher's contacts at the Department of Defense indicated that they would discuss evolved expendable launch vehicles (EELVs) and next-generation launch capabilities. As leaders of the process, the NSC and DPC made statements suggesting that they would not take positions on the details of the proposals. O'Keefe believed that NASA's proposal would win out, both because of the good work Schumacher had done in lining up allies across the executive branch and because the other proposals seemed to him "so much a departure from reality that we [NASA] look downright rational."[21] O'Keefe underestimated the difficulties he would experience in attempting to convince the Deputies of the superiority of NASA's proposal.

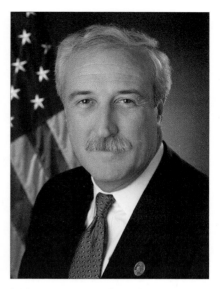

NASA Administrator Sean O'Keefe led the Agency through the Columbia disaster and its aftermath. (NASA)

Considering All Options

As presented at the 25 September Deputies Committee meeting in the White House Situation Room, NASA's space exploration proposal was just one of five

20. Lieutenant Commander Thomas Williams to John Schumacher, email, 15 September 2003, in Mary Kicza to John Schumacher, 16 September 2003, Vision Development emails folder 1 of 3, box 3.

21. Sean O'Keefe to John Schumacher, 18 September 2003, quotation on p. 1, Vision Development emails folder 1 of 3, box 3; Sean O'Keefe to John Schumacher, 22 September 2003, Vision Development emails folder 1 of 3, box 3.

Artist's conception showing various configurations of the Orbital Space Plane, circa 2000. (NASA)

potential options. The format for the meeting required the Deputies to give equally serious consideration to all proposals. Following a short introduction from Stephen Hadley and Margaret Spellings, presenters from OMB, OSTP, NASA, and the CEA gave 5- to 7-minute briefings on their respective proposals, in succession and without intervening discussion.[22] While a couple of the proposals overlapped and shared common goals, the three that attracted the greatest attention represented contrasting visions for the future of space exploration and held starkly different implications in the near term for NASA's plans, programs, and workforce.

The first proposal, OMB's "Mars in our Lifetime," recommended an aggressive astronaut-robotic exploration program that focused on sending humans to Mars as early as 2020. To achieve this ambitious goal, the proposal called for immediately abandoning the Return to Flight effort, terminating the Shuttle Program, ceasing funding for the ISS, and halting the development of the Orbital Space Plane. The OMB plan would give international partners a role in Mars exploration planning to quell criticism of the United States for reneging

---

22. Deputies Committee Meeting, 24 September 2003, pp. 1–2, and Agenda, Deputies Committee 24 September 2003 folder, box 3, Schumacher files.

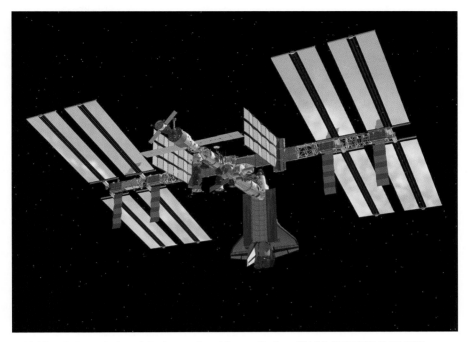

A 2003 artist's rendering of the International Space Station. (NASA JSC2003-E-64498)

on its commitments to the Space Station. The proposal included provisions for developing new launch and crew transfer vehicles and building a lunar base for conducting research on human health conditions in long-duration spaceflight, which the Station otherwise would have supported. It also called for NASA to continue funding Project Prometheus, as well as Earth science and space science research programs. OMB projected that NASA could implement this aggressive plan without an increase in its five-year budget, largely due to the savings accrued from eliminating the ISS and Shuttle Programs. Gaining support in Congress for such a plan, OMB admitted, would be difficult, given the ramifications of canceling existing programs for contractors, workers, and international partners, as well as the broader issue of allowing a decade or more to elapse without a U.S. human presence in space.[23]

A second OMB proposal, called "Human Researchers, Robot Explorers," focused on robotic exploration of the solar system. It differed significantly from the other OMB proposal in that it not only called for maintaining the Space

---

23. Deputies Committee Meeting, "OMB's Proposed Vision #1," 24 September 2003, Deputies Committee 24 September 2003 folder, box 3, Schumacher files; comparison chart, 5 November 2003, Budget Materials folder, Joe Wood files, NASA HRC.

Station, but for improving it so that it could function as a test bed for engineering research and technology development. In terms of space transportation, the proposal recommended the development of a new joint Russian-American, Soyuz-derived vehicle for humans to reach the ISS and contracting with private industry to supply the Station using EELV technology. The Shuttle would remain operational, but only until NASA completed the existing design requirements for the ISS. While continuing support for existing Earth science, space science, and related military space applications programs, the proposal called for eliminating the Orbital Space Plane and slowing the pace of nuclear propulsion research under Project Prometheus to free up funding for other priorities. Like the first OMB proposal, this one would not require an increase in NASA's five-year budget. Yet the proposal held the possibility of attracting criticism for its lack of boldness and for depending on the Russian Soyuz as a foundation for the development of the next generation of space vehicles. Equally problematic, implementing the proposal might require changes to existing laws to allow NASA to engage in a joint technology program with the Russians.[24]

OSTP designed its plan, "Beyond Footsteps—Putting the Moon to Work," to reduce the costs and risks associated with spaceflight. The plan spoke of developing capabilities for human exploration beyond the Moon in the long term, but it focused primarily on robotic exploration of the universe and exploiting lunar resources with robotics. The proposal rested on the assumption that a robotic spaceflight program aimed at exploiting lunar resources would provide the experience necessary for maintaining a robust human spaceflight program at a relatively low cost. Despite the emphasis on the Moon, the proposal recommended against setting timetables or committing to specific destinations. It also called for the continuation of a modest research effort on a new launch vehicle but suggested postponing hardware development until a later date. The plan recommended keeping the Shuttle operational until the completion of the ISS and eliminating Project Prometheus and the Orbital Space Plane, at least in the near term. By transferring money originally committed to Prometheus and the OSP to new lunar robotic missions, the plan would avoid increases in NASA's five-year budget, although it would generate small budget increases several years later.[25]

The Council of Economic Advisors' proposal, titled "Utilizing Markets in Space Exploration," called for NASA to adopt a market-oriented approach,

24. Deputies Committee Meeting, "OMB's Proposed Vision #2," 24 September 2003, Deputies Committee 24 September 2003 folder, box 3, Schumacher files.

25. Deputies Committee Meeting, "Beyond Footsteps," 24 September 2003, Deputies Committee 24 September 2003 folder, box 3, Schumacher files.

using cost-benefit analysis to determine Agency priorities and ceding responsibility for all technical decisions to private contractors. If forced to weigh the
costs and benefits of all missions and programs, CEA predicted, NASA likely
would alter its balance of astronaut and robotic missions and near-Earth and
far-space missions. The plan recommended splitting NASA into two separate groups, with one group responsible for near-Earth activities and the other
focused on activities beyond LEO. While the far-space group would continue to
retain control over missions beyond Earth's orbit, the near-Earth group would
solicit bids for launch vehicles from private industry. The near-Earth group
would not specify which technologies to use to meet its launch requirements. It
would allow contractors to make their own decisions about launch technologies
based on their own analyses of safety, cost, and effectiveness. Under this plan,
NASA likely would retire the Shuttle quickly and then contract with the private
sector for both human and cargo supply to the ISS. With the assumption that
competitive bidding would reduce costs, CEA believed that the proposal would
allow NASA to operate within its current five-year budget. Although privatiza-
tion of flight services might have negative consequences for national security,
the great benefit of the plan, according to CEA, was that it would strengthen
the commercial aerospace industry.[26]

NASA's proposal was by far the most ambitious. Simply titled "Vision
for U.S. Space Exploration," the proposal contained many of the themes the
Decadal Planning Team had developed and advocated a few years earlier. In
terms of strategy, the proposal paralleled DPT in that it called for adopting
a steppingstones approach to move human exploration beyond LEO and for
aggressively integrating human and robotic capabilities. While leaving out the
go-as-you-pay language of DPT, the proposal suggested following a building-
block approach that depended on the development of capabilities rather than
fixed schedules to enable exploration. In terms of objectives, the proposal played
on the themes of American technological leadership and the nation's tradition
of exploration. It also mentioned the potential benefits of the plan for the commercial economy and for the advancement of military technology. Like DPT
and the Grand Challenges that Wes Huntress proposed and Ed Weiler modi-
fied, this plan placed heavy emphasis on science goals, particularly the quest to
understand the origins of the universe and the search for life on other planets.
NASA maintained, furthermore, that this exploration agenda would inspire a
new generation of scientists and engineers and thereby improve the Agency's

---

26. Deputies Committee Meeting, "Utilizing Markets," 24 September 2003, Deputies Committee
      24 September 2003 folder, box 3, Schumacher files.

ability to groom future space explorers while simultaneously strengthening the foundations of the national economy.[27]

NASA envisioned multiple steps in implementing its proposal. The Agency would build on existing programs in the short term, including applying radio-isotope technology developed under Project Prometheus for robotic planetary missions and completing the Orbital Space Plane to ensure safe flights to and from the Space Station. With safety improvements, the Agency would continue to service the Station, and the Station would remain a test bed for medical experimentation to ensure the safety of future human flights beyond LEO. The NASA plan envisioned human missions beyond the Station within the next decade, leaving open the possibility of operations in both high-Earth orbit (HEO) and on the Moon. While the OSP ultimately would replace the Shuttle for human missions to the Space Station, a new heavy-lift launch vehicle would provide support for the Station and for other operations in Earth's neighborhood. The long-term goal of the program, finally, involved both human and robotic exploration of Mars and throughout the solar system.[28]

NASA, in short, took advantage of the opportunity to propose an ambitious agenda that used existing programs and experimental technologies as a foundation for redirecting the civilian space agenda to aggressive human and robotic exploration of the solar system. The Agency peppered its proposal with quotations from President Bush supportive of a bold exploration agenda. Whether the President was willing to spend as much as was needed to support NASA's proposal, however, was another question entirely. Whereas the other agencies all built their proposals to fit into the current budget, NASA developed its cost estimates only after determining its exploration priorities and goals. The Agency asked for an additional $3–$6 billion per year over the subsequent five years. Although the amount, equivalent to annual budget increases of approximately 15 to 35 percent, might have seemed like a large request, NASA was quick to point out that the Agency's expenditures would still represent less than 1 percent of the total annual federal budget. The Apollo program, NASA reminded the Deputies, consumed 3.5 percent of the annual federal budget at its height.[29]

The Deputies were not prepared to choose from among the proposals at the 25 September meeting. Instead, they narrowed down their options by eliminating the CEA proposal and asking OSTP and OMB to work together to

---

27. Deputies Committee Meeting, "NASA Proposal," 24 September 2003, Deputies Committee 24 September 2003 folder, box 3, Schumacher files.

28. Ibid.

29. Deputies Committee Meeting, "NASA Proposal," 24 September 2003, Vision Development emails folder 1 of 3, box 3.

combine their robotics-oriented proposals, respectively "Beyond Footsteps" and "Human Researchers, Robotic Explorers." The Deputies also asked the NSC to develop a hybrid plan that combined elements of all of the proposals.[30]

Refashioning the Proposals

Although the agencies revised their proposals in subsequent weeks to respond to the Deputies' request for more explicit goals and timetables, the most significant change was perhaps the most subtle one—all proposals emphasized the Moon to a greater extent after the 25 September meeting. NASA added several new deadlines to its proposal. The Agency identified 2010 as the date for completing the Orbital Space Plane crew transport vehicle, 2012 as the earliest possible end date for retiring the Shuttle, and 2030 or beyond as the date for the first human-robotic exploration mission beyond Earth's neighborhood. Although NASA's revised proposal more explicitly incorporated the Moon, it left open the possibility of human operations at facilities located at libration points by 2015 as an alternative to the Moon. OMB's refashioned "Mars in Our Lifetime" now identified 2013 as the first possible date for reaching the Moon, 2018 as the approximate date for developing a crewed lunar base, and beyond 2025 as the date for the first human mission to Mars. Other than in terms of destinations and deadlines, the NASA and OMB proposals did not differ in significant respects from the proposals that these two agencies had put forward earlier. OMB remained committed to canceling the Shuttle immediately, halting the development of the OSP, and withdrawing from participation in the Space Station. NASA continued to advocate allowing major scientific questions to set the course and objectives of its human exploration plans and the state of its technological capabilities to set the pace—at a price tag of an additional approximately $3–$6 billion per year. In the language of DPT, NASA remained committed to an exploration program that was capability- and science-driven rather than destination-driven.[31]

The joint OMB-OSTP proposal, named "Opening the Space Frontier," focused heavily on the Moon. From the earlier OMB proposal, this revised plan took the idea of developing a joint U.S.-Russian Soyuz-based crew vehicle for

---

30. Gil Klinger to Klinger et al. [sometimes he sent email to a group of people including himself first], 3 October 2003, Vision Development emails folder 1 of 3, box 3.

31. Klinger to Klinger et al., 3 October 2003, attachments: "NASA Proposal," 7 October (note: the attachment is from 3 October but either accidentally or deliberately was postdated 7 October), and "Mars in Our Lifetime," Vision Development emails folder 1 of 3, box 3; Isakowitz to Schumacher, "Deputies Options Comparisons," 4 October 2003, Vision Development emails folder 1 of 3, box 3.

the near term. It moved away from OMB's recommendation to improve upon the Station and instead advocated reassessing the final configuration of the ISS, halting ISS assembly by 2009, and leaving the ultimate decision on the fate of the Station until 2012, with 2016 as a potential end date of U.S. participation. The central elements of this plan included sending "swarms" of robots to the Moon to determine whether it contained the resources necessary to support a human base. If the robotic prospectors and rovers found water and other essential resources, NASA would establish a permanent human settlement by 2030 that could serve as a foundation for human exploration of the solar system decades in the future. The plan made no explicit reference to Mars or any other location and projected remaining within current budget guidelines.[32]

NSC's hybrid option, called "Continuing Columbia's Journey," was similar to the joint OMB-OSTP proposal in many details. Like the joint OMB-OSTP proposal, the hybrid option had the Moon as its centerpiece. The proposal recommended canceling the Orbital Space Plane immediately, retiring the Shuttle, halting ISS assembly (in 2010 rather than 2009), deferring Project Prometheus, sending robotic precursors to the Moon starting around 2008, beginning extended human missions to the Moon around 2020, and exploiting the resources of the Moon for potential human missions to more remote parts of the solar system in the distant future. Implementing the proposal would require small budget increases over the current five-year NASA budget and additional modest increases in the projected FY 2010 to 2020 budget. Beyond these small budget differences, the hybrid option differed from the OMB-OSTP proposal in that it advocated initiating human missions to the Moon sooner and more explicitly identified human exploration of the solar system as a priority of the space program.[33]

As the participants in the interagency process refined their proposals with input from the 25 September Deputies meeting, Steve Isakowitz quietly began developing plans to implement the vision and reorganize NASA in anticipation of the President's approval.[34] Isakowitz believed that NASA needed a plan that detailed what it should do (both in terms of elaborating on the President's

32. Klinger to Klinger et al., 3 October 2003, attachment, "Opening the Space Frontier," Vision Development emails folder 1 of 3, box 3; Isakowitz to Schumacher, 4 October 2003, "Deputies Options Comparisons," Vision Development emails folder 1 of 3, box 3; comparison chart, 5 November 2003, Budget Materials folder, Joe Wood files.

33. Klinger to Klinger et al., 3 October 2003, attachment, "The Hybrid Option," Vision Development emails folder 1 of 3, box 3; Isakowitz to Schumacher, "Deputies Options Comparisons," 4 October 2003, Vision Development emails folder 1 of 3, box 3; comparison chart, 5 November 2003, Budget Materials folder, Joe Wood files.

34. Isakowitz interview, p. 76; Isakowitz to O'Keefe, 19 October 2003, attachment, "New Exploration Strategy," Vision Development emails folder 1 of 3, box 3.

vision and practical changes in operations) if and when the White House gave the Agency approval for a new exploration strategy.[35] Under Isakowitz's direction, Brant Sponberg, from the Strategic Investments Division in the Office of the Chief Financial Officer, and Gary Martin set up approximately a half dozen small, heavily compartmentalized teams throughout NASA Headquarters to work on different aspects of the plan. Concerned about the potential for leaks to the media, Sponberg and Martin did not explain even to team leaders why they sought such information, and they insisted that team members not discuss their work with any individuals, including supervisors, other than those on their team. The secrecy of the project "create[d] some friction within the Agency," particularly when Isakowitz, Martin, and Sponberg would ask NASA divisions for information without revealing the motivation for their requests.[36]

The compartmentalized teams followed a tight script. Isakowitz and Martin initiated each team with a formal "Task Plan" that provided a schedule and explained the goals, ground rules, and expected deliverables from the activity (a 10-page white paper and a 30-minute PowerPoint presentation in at least one case). In addition to strict instructions to keep the information confidential, the ground rules directed participants to operate under the assumption that the budget would remain stable and to discuss only programs that already had a basis in existing plans, such as the NASA strategic plan, the plans of individual divisions, and reports from the National Academies.[37]

Former DPT and NEXT participants played critical roles on the teams. Jim Garvin led a team that considered science priorities and activities associated with early lunar reconnaissance, while Harley Thronson led three teams, which respectively developed plans for Mars robotic exploration technologies, a telescope technology program to follow the deployment of the James Webb Space Telescope, and science priorities for the big telescopes. Other participants in this process included Lisa Guerra, Orlando Figueroa, NASA Chief Scientist John Grunsfeld, and John Mankins, who led a team that looked at technology tradeoffs. Isakowitz gave the teams approximately three weeks to put together their draft plans to present for initial review. He then gave them another couple of weeks to finalize their reports. By early November 2003, he had all of the white papers and PowerPoint presentations in his possession to do with as he

---

35. Isakowitz interview, p. 76; Isakowitz to O'Keefe, 19 October 2003, attachment, "New Exploration Strategy," Vision Development emails folder 1 of 3, box 3.

36. Isakowitz interview, p. 78.

37. Lisa Guerra to Jim Garvin and Glen Asner, 28 November 2006, attachment, "Task Plan: NASA Science Priorities for Candidate Human/Robotic Space Exploration Pathways, Action from the Space Architect," undated, Printed Emails to Glen Asner and Steve Garber folder, NASA HRC.

saw fit. Regardless of whether this planning episode had any serious impact on the design of the White House vision, it reengaged many of the individuals at NASA who had participated in DPT and served as a critical step in preparing NASA to take charge in the event that the President approved a new exploration strategy.[38]

For the immediate political battle NASA's leadership faced, Isakowitz could not follow such a relatively leisurely schedule. As the NASA teams worked in secret and O'Keefe's senior staff continued to prepare for the next Deputies meeting, Isakowitz became increasingly concerned that NASA lacked an alternative proposal, a Plan B, based more closely on the budget expectations of the White House. Hadley and his colleagues in the White House, Isakowitz sensed, were using the competitive proposal process to wear down NASA and to compel the Agency to accept OMB and OSTP's budget recommendations. Yet NASA officials could not easily tweak their proposal to conform to a more limited budget. In the event of an impasse, with NASA remaining attached to a human exploration plan that would add $28 billion to its five-year budget and the White House refusing to allow a budget increase, the Agency's credibility would suffer. Planning would proceed in this context, but with NASA sitting on the sidelines with little say in the design of its own long-term strategy.[39]

With O'Keefe's consent, Isakowitz drew up a revised strategy in mid-October based on reduced budget expectations and a hastened schedule for replacing the Shuttle. He emailed it on Sunday, 19 October, to Schumacher and O'Keefe, who was then in Russia. With this plan, the Agency indicated for the first time that it was willing to make tradeoffs between competing priorities and to accept the near-term consequences of reducing its commitment to existing programs. Rather than adding a robust human exploration agenda to existing plans and programs as if both were of equal importance, Isakowitz's plan explicitly identified human exploration beyond LEO as the Agency's highest priority and reordered the Agency's near-term agenda accordingly.[40]

While retaining U.S. commitments to international partners and even accelerating research on Project Prometheus, Plan B called for retiring the Shuttle sooner (in 2008 instead of 2012), minimizing assembly flights to the ISS, and

---

38. Thronson interview, pp. 58–61; Jim Garvin to Harley Thronson, 28 November 2006; Jim Garvin to Lisa Guerra, 28 November 2006; Guerra to Garvin and Asner, 28 November 2006, attachment, "NASA Science in the 21st Century: Life in the Cosmos," undated.

39. Isakowitz interview, pp. 75–76.

40. Isakowitz to O'Keefe, 19 October 2003, attachment, "New Exploration Strategy," Vision Development emails folder 1 of 3, box 3; Isakowitz to Kicza, 19 October 2003, Vision Development emails folder 1 of 3, box 3; "New Exploration Strategy," 20 October 2003, presentations folder, Joe Wood files.

ceasing operations in LEO by 2016. In ceding control over LEO to the military and the private sector, NASA would transfer major national programs, including space communications capabilities and the Next Generation Launch Technology (NGLT) program, to other entities. The new proposal also called for redirecting funding for the Orbital Space Plane to a new Crew Exploration Vehicle and for reducing NASA's budget request from a $28 billion increase over five years to a $15.4 billion increase. Plan B more explicitly identified Mars as a destination, with a human circumnavigation of the Red Planet around 2018 and suggested a human return to the Moon by approximately 2015.[41]

Schumacher received a positive response when he gave a verbal preview of the proposal to Klinger and Alexander on 20 October. In an email to O'Keefe, Schumacher explained that the two White House staff members had left him with the impression that Hadley, who was then committed to keeping NASA budget increases to $1.5 to $2 billion per year, might go along with the proposal. This was the first indication O'Keefe had received that Hadley and his colleagues would entertain increases above NASA's current five-year budget. Klinger also told Schumacher that his boss was hoping to reach a final decision on the vision at the next Deputies meeting.[42]

The Wednesday, 29 October, Deputies meeting focused on resolving lingering differences over the defining issues of the vision. Three remaining proposals remained in play—the latest variant of the NASA plan; OMB's "Mars in Our Lifetime"; and the NSC hybrid, which incorporated the joint OMB-OSTP proposal. The Deputies, however, avoided voting on or discussing specific proposals. Hadley and Margaret Spellings structured the meeting around a series of critical questions. The first question, whether Mars should be the driver or just a part of the exploration plan, provoked little disagreement. The group, which included Hadley, O'Keefe, and Marburger, agreed that the goal of a new NASA vision should be to develop infrastructure and capabilities to enable human exploration of the solar system without limiting future exploration plans to a specific destination. The vision, in the consensus view, should not focus narrowly on sending humans to Mars, nor should it incorporate costs or timetables in reaching Mars. The Deputies suggested leaving such decisions to the future,

41. Isakowitz to O'Keefe, 19 October 2003, attachment, "New Exploration Strategy," Vision Development emails folder 1 of 3, box 3; Isakowitz to Kicza, 19 October 2003, Vision Development emails folder 1 of 3, box 3; "New Exploration Strategy," 20 October 2003, presentations folder, Joe Wood files.

42. O'Keefe to Schumacher, 20 October 2003, Vision Development emails folder 1 of 3, box 3.

only after NASA developed the capabilities and infrastructure to enable human exploration of the solar system.[43]

Marburger favored a large-scale vision for the long-term utilization of space. To achieve the necessary political and economic sustainability, he thought a steppingstones approach that married human and robotic exploration (reminiscent of the DPT approach) would be best. Marburger thus believed that Apollo-style "flag-planting" missions would not be sustainable over multiple presidential administrations and decades. He agreed with the notion that space exploration was a "journey, not a race," and favored robust international partnerships.[44]

Marburger viewed Mars as a limiting distraction from a larger plan because the Red Planet was "symbolic of an Apollo-like venture." In his view, planning to send humans to Mars inappropriately "focused resources not on building up a broad capability to go anywhere you wanted, but on a narrow mission that we didn't know how to do at that time." Without the urgency of the 1960s space race to the Moon, Marburger doubted that the American public would tolerate the even higher risk and much higher financial cost of sending astronauts to Mars. Marburger viewed in situ resource utilization as very useful and saw the Moon as an important steppingstone in the exploration of the whole solar system.[45] As mentioned earlier, Marburger's plan was essentially a hybrid of the OMB-OSTP and then NSC options that called for modest increases in the five-year budget.

The decision to de-emphasize the Red Planet, however, did not mean that the Deputies had ruled against all destination-driven conceptions of the vision. A month earlier, the Deputies had asked for more specific goals and timetables, and the agencies responded by placing greater emphasis on the Moon and laying out specific timetables. By considering Mars in isolation at the 29 October meeting, the Deputies bypassed concerns at the working group level about the role of the Moon. Rather than treating Mars and the Moon as competing destinations, the Deputies placed the two in separate categories, with different criteria for what constituted a "destination" based on the expected date of arrival of human explorers. The Deputies would have NASA direct the balance of its resources for the next several decades to reaching the Moon, but the vision would remain non–destination-driven in the minds of the Deputies since the

---

43. The Deputies likely believed that they might avoid the public relations fiasco that had followed the announcement of President George H. W. Bush's Space Exploration Initiative if they excluded cost estimates for reaching Mars (Schumacher notes, Deputies Committee Meeting, 29 October 2003, pp. 1–2, Vision Development emails folder 3 of 3).

44. Marburger interview, pp. 29, 30, 46.

45. Ibid., pp. 32–33, 38.

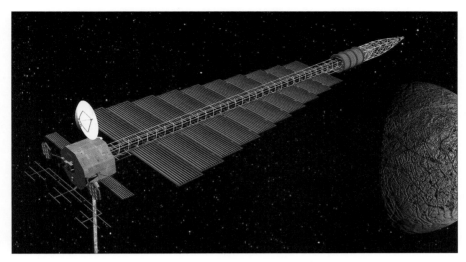

Depiction of the Prometheus Jupiter Icy Moons Orbiter (JIMO) mission to orbit and explore three planet-sized moons of Jupiter: Callisto, Ganymede, and Europa. JIMO would use nuclear electric propulsion to enable the spacecraft to orbit each icy world and perform extensive investigations of their composition, history, and potential for sustaining life. (NASA JPL D2003_0312_S1)

Moon was merely a steppingstone to "other destinations" in the solar system yet to be determined.

Hadley next asked the group whether the first flight of a new Crew Exploration Vehicle (CEV) could be decoupled from the retirement date for the Shuttle and what alternatives existed for servicing Hubble and the ISS after the retirement of the Shuttle. O'Keefe and Richard Armitage from the Department of State argued that allowing a gap between the termination of the Shuttle and the start of CEV operations would force the U.S. into a position of dependence on Russia, limiting the strength of the United States in negotiations on other diplomatic matters and potentially jeopardizing the entire U.S. space program. If the United States could not tolerate a gap in human spaceflight, Hadley responded, then the group needed to choose one of three options: extend the life of the Shuttle, use an upgraded variant of the Soyuz, or accelerate the development of the CEV. O'Keefe reminded Hadley that the NASA proposal, with its emphasis on developing the CEV as soon as possible, was aimed at resolving this problem. The Deputies agreed that the transition between the Shuttle and the CEV, which Armitage referred to as a "technology high-wire act," would be difficult to execute and that they had not reached an entirely satisfactory solution. Most agreed, however, that ceding responsibility to the Russians for transportation to the ISS was not an attractive option. At least at the end of

October 2003, the Deputies opposed allowing a gap between the last flight of the Shuttle and the first flight of the vehicle that would replace it.[46]

The third major question, the role of Project Prometheus, provoked heated discussion as well. Marburger viewed Prometheus as a project for the distant future that could proceed at a slow pace without any serious ramifications for the vision in the near or long term. O'Keefe pushed Prometheus as essential for human exploration beyond LEO and suggested that funding Prometheus would serve as a sign that the administration was serious about long-term exploration. In a follow-up meeting later that day, Hadley expressed concern that Prometheus brought special hazards in an election year given the potential for a general public outcry against the use of nuclear power technologies in space. In a separate follow-up meeting the same day, Marburger argued strongly for tabling the question of the relationship between Prometheus and the vision. He insisted that Prometheus should be allowed to continue on its current path, separate from the human spaceflight program and from discussions concerning the cost and content of the vision.[47]

Hadley declared at the Deputies meeting that the three remaining proposals were not far apart and instructed Gil Klinger to develop a new hybrid option based on the input of the Deputies for review at the next DC meeting, scheduled for the following Thursday, 6 November. Hadley's decision appears somewhat baffling, given just how different the three proposals appeared. With Brett Alexander's assistance, Klinger created a new plan, which he sent to other members of the working group to review.[48] Major issues, nonetheless, remained unresolved. Hadley and the Deputies effectively, if not intentionally, left room for NASA and participants with differing views to fight over the details at the working group level and at the next Deputies Committee meeting.

Although Alexander and Klinger ignored many aspects of the earlier proposals in drafting the new hybrid proposal, the next proposal that Klinger sent to the working group borrowed language on purpose and goals from the NASA proposal. The hybrid proposal, for example, identified the long-term objectives of exploration as strengthening U.S. leadership and innovative capacities and

---

46. Schumacher notes, Deputies Committee Meeting, 29 October 2003, pp. 3–6, Vision Development emails folder 3 of 3; Alexander interview, p. 25; Schumacher interview, p. 29.

47. Schumacher notes, Deputies Committee Meeting, 29 October 2003, pp. 7–12, Vision Development emails folder 3 of 3; Alexander interview, p. 30; "Meeting with Jack Marburger," 29 October 2003, Vision Development emails folder 3 of 3; Klinger to Alexander et al., 29 October 2003, attachment, "Questions for Deputies Committee Discussion," Vision Development emails folder 1 of 3.

48. Klinger to Radzanowski et al., 30 October 2003, attachment, "Hybrid Option - 31 Oct. 03.ppt," Vision Development emails folder 1 of 3, box 3.

answering "ageless questions" about the universe, in the form of Ed Weiler's "Grand Challenges."[49] Other aspects of the proposal reflected the negotiations that had occurred at the 29 October Deputies Committee meeting, such as operating the Shuttle until 2010, continuing the Prometheus project for in-space nuclear propulsion and power, and developing a new CEV with a first flight near the time of the retirement of the Shuttle.

The proposal differed most significantly from the NASA proposal in that it placed strong emphasis on the Moon, with a human return to the Moon between 2010 and 2015. Alexander's boss at OSTP, Bill Jeffrey, was one of the strongest advocates for putting the Moon at the center of the plan. An astronomer by training, Jeffrey promoted a Moon-oriented agenda under the belief that the dusty rock possessed abundant natural resources, including minerals, fuels, and water, that could both support further exploration and contribute to economic growth on Earth.[50] Although they acknowledged OSTP's support for the Moon and believed that most of the Deputies saw the Moon as a logical starting point for human space exploration, Klinger and Alexander took credit for inserting the Moon into the vision in such a fashion that it no longer held the potential to serve as a point of contention at either the Deputies or working group levels. They dismissed as either impractical or politically unacceptable the alternative initial human exploration steppingstones that NASA put forth, including a Mars circumnavigation trip and human-tended astronomical observatories at libration points.[51]

Klinger and Alexander did not fret about the Moon decision. It was a simple and obvious choice for Klinger, at least. "There was no way in my judgment, and [on] God's green Earth," Klinger later declared, "[that] we were going to be able to send human beings to Mars until we first demonstrated that we could safely get them up and back to some other destination, and the only destination that made sense was the Moon."[52] Contradicting the conclusions of DPT, as well as the basic NASA starting position, the two staffers put together a proposal that focused narrowly on a single destination. Steve Isakowitz, for one, was

---

49. The document listed three of Weiler's Grand Challenges: "Where do we come from? Are we alone? Where are we going?" See chapter 3 for a discussion of the Grand Challenges of both Wes Huntress and Ed Weiler. Klinger to Radzanowski et al., 30 October, attachment, "Hybrid Option – 31 Oct. 3.ppt," Vision Development emails folder 1 of 3, box 3; Alexander interview, p. 31.

50. Isakowitz interview, p. 62; Schumacher interview, p. 24. Schumacher was unsure whether Jeffrey was acting on his own or following Richard Russell or Marburger's lead on the Moon and lunar resources.

51. Alexander interview, p. 32; Klinger, pp. 49–51.

52. Klinger interview, p. 49.

"disturbed" by what he perceived as the "Moon or bust" attitude of the White House staff members who were treating the Moon as an end in itself.[53]

The new hybrid proposal left a great deal of slack between deadlines for reaching milestones, such as between 2010 and 2020 for the development of a lunar lander and lunar habitation infrastructure. The proposal, most importantly, defined the key remaining unresolved issues, such as what technology to use to reach the ISS after the retirement of the Shuttle, the role of international partners in the vision, and the length and desirability of the gap between the Shuttle's retirement and the deployment of a new ISS transportation system.[54] Of the three potential budget options, one failed to give definitive numbers. The authors identified it as a "schedule driven" approach in which spending fluctuated to ensure adherence to a predetermined timetable. The second option called for no change in the current five-year budget, between fiscal years 2005 and 2009, while the third included incremental increases to total $7 billion between fiscal years 2005 and 2009 and then $2 billion every year thereafter.[55]

The negotiation process between NASA (Isakowitz), OMB (Radzanowski), and the NSC (Klinger) over the hybrid proposal produced mostly minor changes in emphasis. Isakowitz pushed for making Mars a more prominent part of the proposal and reducing the emphasis on the Moon. The working group, nonetheless, kept the Moon as the centerpiece of the proposal, although only as the first major steppingstone in a grander, long-term exploration vision.[56]

The proposal mentioned Mars solely as a potential destination among others to visit throughout the solar system and extended even further the possibilities of reaching certain milestones. Klinger's initial 31 October hybrid proposal, for example, called for introducing the new CEV between 2010 and 2015. The next version of the proposal to the Deputies broadened the range of dates for introduction of the CEV to between 2009 and 2020. Whereas the first proposal merely left the decision of where to go after the Moon until 2025, the version for the 6 November Deputies meeting incorporated discussion of robotic precursor exploration of the solar system between 2004 and 2020 and demonstrations of space nuclear fission propulsion as early as 2012. The early November proposal

53. Isakowitz to O'Keefe and Schumacher, 5 November 2003, Vision Development emails folder 1 of 3, box 3.

54. Klinger to Radzanowski et al., 30 October 2003, attachment, "Hybrid Option – 31 Oct. 3.ppt," Vision Development emails folder 1 of 3, box 3.

55. Ibid.

56. Klinger to Boyd et al., 5 November 2003, attachment, "Hybrid Option (post meeting) 04 Nov.ppt," Vision Development emails folder 1 of 3, box 3; Schumacher to O'Keefe, "Deps," 5 November 2003, attachment, "Hybrid option (post-meeting) –04 Nov2003.ppt," Vision Development emails folder 1 of 3, box 3.

also focused more attention on the question of whether a gap in transportation to the ISS was acceptable and what to do with the Station if NASA did not complete it by 2010, the year designated for the retirement of the Shuttle.[57]

Although Isakowitz, Klinger, and all other members of the working group put their support behind the NSC hybrid in the hopes of bringing the Deputies' process to a positive conclusion, they left a few critical issues for the Deputies to decide. The two most important issues, the adjustment to the five-year budget required for implementation and the acceptable delay in U.S. human space operations between the last Shuttle flight and first CEV flight, were initially joined in a set of charts that laid out various budget options, ranging from a freeze to the fiscal year 2005 budget, which would result in a $4.9 billion five-year budget decrease, to an increase of $18.2 billion, which would eliminate the gap between the last flight of the Shuttle and the first flight of the CEV.[58]

A day before the 6 November Deputies meeting, OMB altered both the calculations and the design of the budget charts so that the relationship between the gap and the budget was not so apparent. Isakowitz described OMB's last-minute budget changes as "mischief," while Schumacher informed O'Keefe that OMB was now pushing for restricting budget increases as much as possible.[59]

NASA thought it had gained an advantage in budget discussions when, earlier in the process, the working group anointed Steve Isakowitz "the official scorekeeper" for determining the costs of all proposals and options. Isakowitz viewed this responsibility as a "critical strategic advantage," since it allowed him to ensure the uniformity of the assumptions entering into the costing of the proposals. However, although this gave Isakowitz the ability to ensure the fairness of the process,[60] the determination of what assumptions he would use to cost the proposals remained beyond the control of NASA, much to the chagrin of O'Keefe and Isakowitz. As long as White House officials remained committed to restricting the human exploration budget, Isakowitz's accounting responsibilities gave NASA little leverage.

57. Klinger to Boyd et al., 5 November 2003, attachment, "Hybrid Option (post meeting) 04 Nov.ppt," Vision Development emails folder 1 of 3, box 3; Schumacher to O'Keefe, "Deps," 5 November 2003, attachment, "Hybrid option (post-meeting) –04 Nov2003.ppt," Vision Development emails folder 1 of 3, box 3.

58. Klinger to Radzanowski, "Re: Final Updated Budget Options," 5 November 2003, Vision Development emails folder 1 of 3, box 3; Isakowitz to Radzanowski, "FWD: Final Updated Options," 5 November 2003, Vision Development emails folder 1 of 3, box 3.

59. Schumacher to O'Keefe, "Deps," attachment, "Hybrid option (post-meeting) –04 Nov2003. ppt," 5 November 2003, Vision Development emails folder 1 of 3, box 3.

60. Isakowitz interview, p. 72.

Joel Kaplan, the newly appointed deputy director for OMB, compelled his subordinates to abandon the method for calculating budget alternatives that they had followed for the previous several months. Rather than using the President's budget for FY 2004 to FY 2009 as the base for estimating the increases that the various alternative scenarios would require, Kaplan insisted that OMB should start with the assumption that the budget would remain frozen at the FY 2004 level in calculating future-year budgets with investments for the Vision included. Using the OMB method, the budget request for NASA's initial proposal grew from $18 billion to $27 billion. All proposals, for that matter, increased $8.7 billion between FY 2005 and FY 2009 as a consequence of OMB's new methodology.[61]

Moon, Mars, the Gap, and the Budget

The 6 November 2003 Deputies meeting started out on a positive note. Gil Klinger presented the briefing that all members of the working group had approved despite the continued misgivings of some about the Shuttle-CEV gap and the budget calculations. Remarking that the proposal laid out a good range of options for the vision, Marburger praised Klinger for giving an excellent presentation and thanked O'Keefe for bringing the process closer to a conclusion. Representatives from DOD and the intelligence community both consented to the proposal as Klinger had presented it. Joel Kaplan from OMB indicated that he was comfortable with it and thought that it was ready for presentation to the President.

The tone of the meeting changed, however, once Armitage homed in on the major issues that remained in dispute: the Shuttle-CEV gap and the budget. From a diplomatic-relations perspective, Armitage thought it was critical for the United States to have continuous human operations in space. He argued strongly against allowing a gap between Shuttle and CEV operations, but he also expressed concern about attempting to limit the gap without adequate resources. A planned gap meant dependence on the Russians. Putting too much pressure on NASA to speed the completion of both the ISS and the development of the CEV without adequate resources, Armitage implied, would introduce

---

61. Isakowitz to O'Keefe, "Re: Deps," 6 November 2003, Schumacher files, box 3, Vision Development emails folder 3 of 3, box 3; Schumacher to O'Keefe, "Deps," 5 November 2003, Vision Development emails folder 1 of 3, box 3.

technical problems and ultimately cost more than an adequately funded transition program.[62]

Armitage's comments provided an opening for O'Keefe to explain the cost and scheduling implications of the various options for retiring the Shuttle and completing the Station. O'Keefe suggested that it would be misguided to design a new vision with the implicit assumption that retirement of the Shuttle might slip to a later date. The CAIB Report had recommended retiring the Shuttle as soon as possible and not later than 2010. If NASA chose to fly the Shuttle past 2010, the CAIB recommended complete recertification of the entire vehicle and all its components—a costly process that would take several years to finish and require about $3 billion. Investing adequate resources to complete the CEV in a timely fashion, for this reason, would save money in the long run. Such an achievement also would have a positive impact on employee morale. For these reasons, O'Keefe argued that many of the alternatives and options that the Deputies were considering could not "get you to the vision." They could not even "be made to sound plausible," he explained, hinting that Congress, NASA contractors, and NASA employees would balk if the administration put forth a proposal that did not conform to conventional expectations in terms of schedule and cost for a long-term human exploration program.[63]

Hadley warned O'Keefe not to force "false choices" on the President by suggesting that the only viable approach to the vision was to design it with a seamless transition between Shuttle and CEV operations in 2010. Although he and Armitage worried that such an approach would make a new vision less ambitious, O'Keefe was willing to entertain Hadley's suggestion to extend Shuttle operations and delay the completion of the CEV until 2014. Yet O'Keefe still questioned the logic of saddling NASA and the taxpayers with the certain costs of recertifying the Shuttle for operations past 2010 and the potential costs in terms of human life and national prestige of continuing to operate a space vehicle long past its prime. The real long-term costs of the CEV, O'Keefe suggested, would not change whether NASA followed a delayed or an accelerated development plan. The only clearly known and avoidable cost was the cost of Shuttle recertification.[64]

None of the Deputies discussed Shuttle retirement and CEV development with reference to electoral or budgetary cycles. In all NASA and OMB

62. Schumacher notes, Deputies meeting on exploration, 6 November 2003, pp. 1–2, Vision Development emails folder 3 of 3; Schumacher interview, p. 29.

63. Schumacher notes, Deputies meeting on exploration, 6 November 2003, pp. 2–3, Vision Development emails folder 3 of 3.

64. Schumacher notes, Deputies meeting on exploration, 6 November 2003, pp. 2–5, Vision Development emails folder 3 of 3.

projections, spreading CEV development over 10 years, rather than 6, resulted in only modest changes to the President's five-year budget for NASA. By putting forth a "budget neutral" vision (whether real or fictional) and delaying the completion of the CEV, Hadley and his counterparts may have been attempting to protect the White House from the political battles that a vision with an immediate price tag might provoke. Such an approach may have amounted to merely passing the responsibility for paying the true costs of CEV development to the next five-year budget cycle and to the next presidential administration. Regardless of whether Hadley and some of the other Deputies operated from such a perspective, O'Keefe appears to have made his case for the vision without great sensitivity to the political implications of a large budget increase for the Bush administration. O'Keefe sought to develop momentum for the vision by funding major priorities up front and moving quickly to open a new era of human spaceflight.

The 6 November meeting ended without resolution of the issues O'Keefe and Armitage had raised. O'Keefe and Hadley agreed that they needed additional information on the cost and advisability of various options before reaching a final decision. The Deputies' options for dealing with the transition from the Shuttle to the CEV, as Schumacher understood it, included retiring the Shuttle and beginning CEV operations in 2010, recertifying the Shuttle and continuing to operate it until the CEV became operational, or retiring the Shuttle in 2010 and then relying on the Russians until the CEV reached operational status. The NSC rejected consideration of a fourth option—to build a Soyuz-derived vehicle in the United States with the assistance of the Russians.[65]

As the basic outline of the vision solidified, the haggling over the budget and the Shuttle-CEV gap intensified. O'Keefe continued to lobby behind closed doors for an accelerated program that would allow for a continuous U.S. human presence in space. While pushing firmly for explaining the various budget and Shuttle transition options to President Bush, he worked with the other Deputies to set the stage for a meeting with the President. O'Keefe understood that the group could not put forth an incomplete proposal that left a long list of issues for the President to decide. At the same time, he believed that it was only fair for the Deputies to provide the President with clear information on the costs of various policy options and the different methods for calculating the budget, specifically whether to freeze the NASA budget for calculating the vision at the FY 2004 appropriation level or to use the FY 2004 President's budget

65. Schumacher notes, Deputies meeting on exploration, 6 November 2003, pp. 5–6, Vision Development emails folder 3 of 3; Schumacher notes, post-Deps meeting, 6 November 2003, Vision Development emails folder 3 of 3.

request, with increases in fiscal years 2005–2009, as the baseline for calculating the cost of the vision. In a 13 November meeting, the Deputies attempted to dissuade O'Keefe from raising these issues with the President. Perhaps because of O'Keefe's forcefulness, they relented and appeared to have consented to allow him a few words on the budget in the final meeting with President Bush.[66]

Over the next few weeks, O'Keefe held several teleconferences and private meetings with key officials in the White House, including Andrew Card, Josh Bolten, Karl Rove, and Steve Hadley.[67] Although no records exist of meetings between O'Keefe and his old Department of Defense colleague, Vice President Dick Cheney, the two likely had several private conversations about the vision. On Friday, 14 November, for example, O'Keefe asked Isakowitz to prepare charts for a meeting the following week with Cheney. O'Keefe likely hoped that he could convince his old DOD colleague to use his considerable political heft in the White House to ensure that the vision reflected NASA's priorities and budgetary expectations. For the overall tone of the vision, as well as for ongoing budget discussions with OMB, the NSC, and the White House, O'Keefe's personal relationship with Cheney appears to have made little difference.[68] O'Keefe also made little headway with Josh Bolten and Andrew Card.

"Black Tuesday," as Isakowitz later called it, came two days before Thanksgiving, on 25 November, while O'Keefe and his senior staff were at a retreat in Ashburn, Virginia. Andrew Card called O'Keefe that day to let him know that the negotiation process was over. A meeting with the President would not take place unless NASA developed a plan for reorganizing the Agency based on budget calculations that fit within OMB's guidance, which amounted to

---

66. Schumacher notes, Deputies meeting, 13 November 2003, Vision Development emails folder 2 of 3.

67. Sean O'Keefe's hard-copy calendar for November and December 2003, [O'Keefe] Calendar 5/03-3/04 in O'Keefe files, NASA HRC.

68. O'Keefe to Schumacher, 13 November 2003, Vision Development emails folder 2 of 3, box 3; Vision Development emails folder 2 of 3, box 3; Schumacher to Isakowitz, 14 November 2003, Vision Development emails folder 2 of 3, box 3; Isakowitz to Schumacher, 14 November 2003, Vision Development emails folder 2 of 3, box 3; Isakowitz to Retha Whewell, 18 November 2003, Vision Development emails folder 2 of 3, box 3; Pastorek to O'Keefe, 25 November 2003, Vision Development emails folder 2 of 3, box 3; Schumacher interview, p. 27. Alexander claimed that "[t]he Vice President's office never went to bat for Sean." Alexander interview, p. 46. O'Keefe had known Cheney since the first Bush administration, when Cheney served as Secretary of Defense and O'Keefe served as Secretary of the Navy. O'Keefe's calendar did not show any scheduled meetings or phone conversations with Cheney in November or December. Mentions of O'Keefe's intention to schedule a meeting with the Vice President dropped off on 19 November. O'Keefe appears not to have met with Cheney, although it is likely that the two had an unscheduled phone conversation. According to John Schumacher, O'Keefe's Chief of Staff, O'Keefe and Cheney spoke regularly and O'Keefe occasionally spoke with President Bush.

$1 billion over the President's FY 2004–2009 budget request, but approximately $11 billion if the Agency used the frozen FY 2004 appropriations as the baseline for calculating the budget. The interagency process, Schumacher scribbled in his notes that day, had left NASA with a coherent, consensus vision, "but not even a bake sale to pay for it." Although initially upset about the call, even depressed, O'Keefe and the members of his inner circle resolved to meet the challenge by setting to work identifying which programs to cut or scale back in order to initiate the vision with much less money than they thought the Agency needed. Only one person, Paul Pastorek, an old friend of O'Keefe's who was serving as NASA's General Counsel, attempted to persuade him to take a hard stand and to continue to fight for a larger budget increase. O'Keefe opted instead to give Isakowitz time to generate a new plan before gambling on the entire vision.[69]

Isakowitz pulled a small group of people away from their Thanksgiving celebrations, including Liam Sarsfield, the NASA Deputy Chief Engineer for Programs; John Grunsfeld, the NASA Chief Scientist and an astronaut; and Brant Sponberg, from the Strategic Investments Division in the Office of the Chief Financial Officer, to crunch numbers and to refine the Plan B he had developed a month earlier with the assistance of some of these individuals and others from NASA Headquarters. Accepting that they would have to make do with just $11 billion reprogrammed from the current five-year budget—$17 billion less than they initially sought—Isakowitz and his team identified ways to lower the total cost of NASA's plan by reducing the Agency's immediate investment in the CEV and cutting the science budget.

Isakowitz found savings by reconfiguring the plan for the development of the CEV based on the concept of "spiral development," which was then growing in

---

69. Quotation from Schumacher meeting notes, 25 November 2003, Vision Development emails folder 2 of 3, box 3. Pastorek sent O'Keefe an email that evening urging him to fight for the initial NASA budget request. Schumacher and Associate Administrator for Space Flight Bill Readdy sent emails the following morning, giving O'Keefe the opposite advice: run with the policy the White House was offering. Readdy and Schumacher agreed that Andrew Card was not bluffing and that it was time for NASA to take ownership of the vision. Isakowitz interview, pp. 80–83; Schumacher interview, p. 28; Alexander interview, p. 41; Readdy to O'Keefe, 26 November 2003, Vision Development emails folder 2 of 3, box 3; Pastorek to O'Keefe, 25 November 2003, Vision Development emails folder 2 of 3, box 3; Schumacher to O'Keefe, 26 November 2003, Vision Development emails folder 2 of 3, box 3.

For more general background on the budget negotiations at this time, see W. Henry Lambright, *Why Mars: NASA and the Politics of Space Exploration* (Baltimore: Johns Hopkins University Press, 2014), pp. 186–187. See also Steven J. Dick, "Appendix: The Decision to Cancel the Hubble Space Telescope Servicing Mission 4 (and Its Reversal)" in *Hubble's Legacy: Reflections by Those Who Dreamed It, Built It, and Observed the Universe with It,* ed. Roger D. Launius and David H. DeVorkin (Washington, DC: Smithsonian Institution Scholarly Press, 2014), pp. 158–160, although this source's discussion of budget negotiations at this time is obviously more focused on the Hubble Servicing Mission.

popularity in the defense community as a method for funding and managing major, long-term, high-risk technological projects. Spiral development of the CEV would follow several phases, defined in terms of prototype vehicles, with the earliest prototypes consisting of vehicles with the fewest components necessary to achieve the most critical functions. The vehicle would gain new functions with each new iteration, or successive spiral. The approach would allow NASA to start designing and testing the CEV with a relatively small investment. The central disadvantage, as Isakowitz readily admitted, was that it would increase the total cost of a system over the life of its development.[70]

As for the science programs, Isakowitz knew that any cuts would elicit serious opposition from the scientific community, so he exercised great caution in identifying programs to eliminate. Among the most widely circulated products of the Thanksgiving number-crunching of Isakowitz and his colleagues was what would become known as the "sand chart," which showed the projected growth and redistribution of NASA's budget across major program areas (Shuttle, human exploration, aeronautics, science, etc.) from fiscal years 2004 through 2020.[71]

Pleased with the work of Isakowitz's group, O'Keefe convened a meeting of his Executive Council on 2 December 2003 to gauge the level of support for the plan from senior NASA officials. Some officials, such as Office of Safety and Mission Assurance Associate Administrator Bryan O'Conner, expressed concern about the potential impact of the draft vision on existing programs, particularly the Shuttle and Space Station. Nonetheless, broad agreement that the emerging vision represented an improvement over existing Agency plans stifled serious criticism. Associate Administrator for Aerospace Technology Victor Lebacqz, for example, expressed support for the plan, even though he recognized that it would force the Agency to scale back aeronautics research and cut his division in half. Ghassem Asrar, similarly, made only a mild effort

---

70. Isakowitz to O'Keefe, 5 November 2003, Vision Development emails folder 1 of 3, box 3; Schumacher interview, p. 30. Spiral development, in other words, was politically advantageous in the short term, and perhaps a wise choice from a technological perspective, but it was not necessarily the most cost-effective approach in the long term. On the other hand, spiral development also provided more opportunities for a funding agency to test, evaluate, and even cancel a project before the sunk costs became too great. Admiral Craig E. Steidle, an advocate of spiral development who would serve as NASA's first Associate Administrator for Exploration following the announcement of the vision, may have inspired Isakowitz to adopt the spiral development approach. Isakowitz and other members of NASA's leadership team had been in contact with Steidle since the summer. According to Gary Martin, he and other NASA officials interviewed Steidle in late summer or early fall 2003. See Martin interview, 16 February 2006, p. 19.

71. Isakowitz interview, pp. 82–86. The "sand chart" is reproduced in appendix D-4.

to make his division, Earth Sciences, an integral part of the Agency's future plans. Ed Weiler, then head of the Office of Space Science, gave what DPT participants would consider the punch line of the meeting when he remarked that it would be hard for him not to support the vision, given that he had been working on it for five years.[72]

Consent to the vision and to the budget cuts that would go with it, in O'Keefe's mind, meant that NASA senior managers had relinquished all rights to complain about it in the future or to try to alter the budget distribution. In presenting the NASA budget request the next day, 3 December, to OMB budget examiners, Isakowitz boasted that the Agency's senior leadership team supported the vision and that O'Keefe anticipated little resistance from them after the announcement of the vision.[73]

With the budget and the basic outlines of the plan resolved, Isakowitz, Klinger, and their counterparts at the working group level shifted their attention to figuring out the steps to be taken should the President approve the plan. They focused on the logistics and details of the President's vision announcement and developed strategies for dealing with Congress, international partners, and the media. O'Keefe also stepped up preparations for the Agency's post-announcement public relations campaign. For example, Glenn Mahone, Assistant Administrator for the NASA Office of Public Affairs, provided O'Keefe with a seven-page bullet-point list of themes and rationales for the Agency to use in justifying the vision to the public and in preparing its workforce for what would be the most significant transition in decades. The individuals involved at NASA, the White House, and other agencies kept all aspects of their activities secret, fearing that the President might kill the entire proposal if word of it leaked to the media.[74]

A subtle change in the tone of the presentation for the President occurred after Sean O'Keefe met on 9 December with Karl Rove, an unidentified individual

---

72. Schumacher notes, "Vision, 2 p.m. readout w/Ent. AAs Excmte.," 2 December 2003, Vision Development emails folder 2 of 3, box 3.

73. Isakowitz to O'Keefe, 3 December 2003, Vision Development emails folder 2 of 3, box 3.

74. Others involved in discussions of the Agency's public relations rollout for the vision included regular attendees of O'Keefe's "strategery meetings," such as Fred Gregory, Schumacher, Mahone, Isakowitz, Legislative Affairs Assistant Administrator Lee Forsgren, and External Relations Assistant Administrator Mike O'Brien (Mahone to O'Keefe, 5 December 2003, Vision Development emails folder 2 of 3; Klinger to Steve Isakowitz et al., "Subject: Next Steps on Vision," 2 December 2003, Vision Development emails folder 2 of 3, box 3). The term "strategery" was coined for an October 2000 *Saturday Night Live* television sketch lampooning then-presidential candidate George W. Bush's tendency to stumble verbally. After he won the presidency, Bush administration aides began to embrace the term. See, for example, *https://en.oxforddictionaries.com/definition/strategery* (accessed 19 July 2018).

from the Vice President's office, and several other senior White House officials who had been involved all along, including Miers, Hadley, Spellings, Marburger, and Joel Kaplan. The participants discussed the role of international partners in the vision and the importance of choosing the right venue for the President's announcement. They also set Monday, 15 December, as the day President Bush would get a copy of the presentation and Thursday, 18 December, as the day the formal presidential decision meeting would occur. O'Keefe scored a minor victory in this meeting in that he received approval to insert a version of Gary Martin's steppingstones chart, derived from his days with DPT, into the presentation for President Bush.[75]

From his position at OMB, Dave Radzanowski played a key role in coordinating the development of the VSE policy. (NASA)

While seemingly innocuous, the chart promoted Mars from merely being one of several potential destinations after the Moon to becoming NASA's central ambition after developing surface operation capabilities on the Moon. NASA went even further with another chart listing specific milestones, including dates for the Mars Robotic Sample Return (2013) and a Mars circumnavigation by humans (2020 or later). Dave Radzanowski of OMB protested the shift of emphasis. He and his colleagues argued that the presentation should not include any mention of Mars, even the Mars Robotic Sample Return project that was already on the science program's priority list. While Radzanowski was not necessarily opposed to a future Mars mission, he and his colleagues opposed including any slides that might imply that the Deputies had agreed to Mars as a milestone.[76]

Isakowitz objected strenuously. He worried that removing references to Mars would make "the vision sound like a 'moon or bust' stovepipe." As he explained to Klinger, Schumacher, Radzanowski, and Brett Alexander, "The vision does not

---

75. O'Keefe's calendar [listed incorrectly as Tuesday, 8 December]; Schumacher notes, "Follow up to SOK [Sean O'Keefe] mtg," 9 December 2003, Vision Development emails folder 2 of 3, box 3; "MTW SOK re Vision," 9 December 2003, Vision Development emails folder 2 of 3, box 3; "Briefing for the President," 19 December 2003, Brett Alexander folder, NASA HRC; Klinger to Isakowitz, 4 December 2003, Vision Development emails folder 2 of 3.

76. Radzanowski to Klinger et al., 10 December 2003, attachment, "SOKmtgRove12-9-032.ppt," Vision Development emails folder 2 of 3, box 3.

ban us from going to Mars. Indeed, the vision calls for exploration of the solar system, and last I checked, it [the solar system] includes Mars."[77] Klinger initially tried to appease both sides, telling them that he agreed with Radzanowski that the presentation should remove any sense that the Deputies had agreed to specific milestones for Mars exploration, but that he also agreed with Isakowitz that they needed "to ensure that the vision is not perceived as lunar centric."[78] Klinger took both perspectives into account and set to work compiling a nearly final version of the presentation for the President that ultimately reflected the concepts O'Keefe, Rove, Miers, Hadley, Spellings, Marburger, and Kaplan had agreed upon the prior day and separate changes that Hadley had directed Klinger to incorporate that day, 10 December.[79]

Presidential Decision

The consensus vision to be presented to the President reflected a compromise between the various interagency participants, particularly NASA, OMB, and OSTP. Individuals at the working group level worked hard to limit the differences of the organizations they represented to provide the President with a clear path for reaching a decision on the future of NASA. By the day of the briefing, only three basic issues remained for the President to decide:

- Was the plan under consideration the right policy for the U.S. space program at that time?
- Should the United States invite other nations to participate in future exploration activities?
- What type of commission should the White House create to preempt congressional calls for the creation of an independent commission to assess NASA's exploration program?

At the instruction of Stephen Hadley, Klinger removed the chart with timetables for Mars sample return and Mars circumnavigation flights that OMB found objectionable. Yet a Mars destination remained in the charts as NASA had insisted.

The meeting began at approximately 2:20 p.m. on 19 December 2003 in the Roosevelt Room at the White House. The President sat in the middle of one of the long sides of the rectangular table, with Cheney and I. Lewis "Scooter" Libby to his right and Card, Miers, and Rove to his left. Margaret Spellings sat

---

77. Isakowitz to Klinger et al., 10 December 2003, Vision Development emails folder 2 of 3, box 3.
78. Klinger to Isakowitz, 10 December 2003, Vision Development emails folder 2 of 3, box 3.
79. Klinger to Alexander, 10 December 2003, Vision Development emails folder 2 of 3, box 3.

directly across from the President, flanked on her left by Hadley, Klinger, and Marburger. On her right sat O'Keefe and Kaplan. White House press secretary Scott McClellan, speechwriter Mike Gerson, and Legislative Affairs Assistant David Hobbs also sat at the table. Among the individuals in chairs surrounding the table were Radzanowski, Montgomery, Russell, Miller, Schacht, and Schumacher. Gil Klinger had the honor of briefing the President and his senior staff on the core elements of the plan and the issues that remained for the President to decide.[80]

After Margaret Spellings briefly explained the purpose of the meeting, Gil Klinger began his presentation with a basic outline and a list of decisions that remained for the President to make. He then jumped into the reasons for the vision. The fourth slide asked, "Why do we need a new Vision?" The slide highlighted criticisms from the Columbia Accident Investigation Board and Congress that placed blame on NASA for lacking a vision or overarching strategic plan to guide its efforts and for failing to develop a new transportation system. The slide quoted the CAIB Report's claim that "previous attempts to develop a replacement vehicle for the aging Shuttle represent a failure of national leadership."[81] The points on leadership and past failures caught the attention of President Bush, a Harvard M.B.A. and former businessperson, who interjected a few questions about NASA's lack of vision and failure to replace the Shuttle.[82] As Klinger presented it, the new strategy would be an opportunity for President Bush to correct the failings of past leaders who lacked awareness of the importance of grand strategic plans, or visions, for uniting an organization, imposing discipline on its leaders, and providing its employees with a sense of purpose.

A steppingstones chart appeared prominently as the first graphic five pages into the presentation. The various drafts of the briefing in the weeks before the meeting with the President had contained several different versions of the steppingstones chart, reflecting ongoing discussions and debates about the wording of each bullet point and idea. Almost all of the charts presented a similar visual of progress from the ISS in LEO to 30-day missions to the Moon and 365-day missions to Mars. Significantly, only the final version for presentation to the President identified the Moon as a "Stepping Stone to Exploration." The visual

80. Schumacher notes from Presidential Decision on Space Exploration, 18 December 2003, Vision Development emails folder 2 of 3; Alexander interview, p. 44; Radzanowski, Record, "Policy Time on NASA," 20 December 2003, Radzanowski files. The meeting occurred one day after the date the Deputies initially set for it, most likely to accommodate the President's schedule.

81. "Briefing for the President: Future U.S. Space Exploration, Alternative Visions, Key Elements, and Issues for Decision," 19 December 2003, Vision Development emails folder 2 of 3.

82. Robert Draper, *Dead Certain: The Presidency of George W. Bush* (New York: Free Press, 2007).

allowed for the possibility that operations on the Moon could contribute to further exploration by providing opportunities for testing new technologies and operational concepts. Yet, taken as a whole, the goals for the Moon—30-day missions, development of surface operations capabilities, investigations of lunar resources, and the Moon as a steppingstone—implied that the Moon was not the ultimate goal of the plan or the termination point for human exploration.[83]

The President seemed to grasp the value of a steppingstones approach while also understanding the "political importance of having an objective beyond the Moon." According to Marburger, policy-makers "really did have to have an objective, and the president nodded at Mars." While Marburger disliked a focus on Mars that excluded other goals, he knew Mars was a convenient shorthand because the solar system–wide vision he preferred "was just too complicated."[84]

The President worried that the plan might not generate much public enthusiasm if it focused solely on returning to the Moon or appeared to be a jobs program for the city of Houston. He wanted the plan to identify specific targets for human exploration beyond the Moon, particularly Mars. Bush insisted that the final proposal clearly indicate that the plan would involve extended stays on the Moon to demonstrate capabilities and mine resources in preparation for human exploration of Mars and then other destinations. Questions about opportunities for making the plan seem more exciting evoked a mention of the possibility of launching from the Moon. President Bush believed that public support would be forthcoming if concepts like launching from the Moon for humans to venture to other destinations were made more explicit and conveyed visually.[85]

Informal notes from the meeting also indicate that someone, perhaps the President, may have questioned whether NASA should continue its space science and Earth science programs. The questions arose in the course of a discussion in the middle of the presentation on the budget and programmatic changes required to implement the VSE. The slides called for $1 billion in additional funding over five years and $11 billion reprogrammed from other activities. Most of the reprogrammed funding would come from retiring the Shuttle in 2010. The second source of reprogrammed funds the slide suggested would come from NASA's space science and Earth science programs. Next to a bullet point indicating that the space and Earth sciences programs would bear the brunt of the costs of restructuring NASA to implement the VSE, Schumacher's notes indicated that the "[question] is why fund at all—why not zero?" While

83. "Briefing for the President," p. 5; Radzanowski, Record, 22 December 2003, p. 2.
84. Marburger interview, pp. 39, 33, 45.
85. Schumacher interview, p. 27; Isakowitz interview, pp. 88–89; Alexander interview, p. 41; "Briefing for the President," p. 7; Schumacher notes.

only suggestive of the administration's thoughts regarding NASA's science programs, the initial bullet itself that read "Principal impact on Space Science and Earth Science components of existing NASA program," also implies that the White House viewed the scientific programs that were unrelated to the VSE as a low priority for NASA.[86]

The most contentious issue that emerged in the meeting was the gap between the retirement of the Shuttle and the first flight of a new CEV. The President questioned the wisdom of relying on Russian Soyuz vehicles for flights to the Space Station, but not for the same reasons O'Keefe had articulated earlier.[87] O'Keefe worried that dependence on Russia for access to space would put NASA and even the U.S. government in a vulnerable position diplomatically. While the President asked whether dependence on Russia would harm morale, presumably within the space program, he did not indicate any concerns about maintaining positive relations with Russia in the future. He assumed that the problem with dependence was that the Russian space program was less concerned about safety than NASA. Andrew Card, however, told the President that the Russian space program had a better safety record than NASA. Reminded of the failures of the American space program in the past year, the President quickly dropped his concerns about dependence.[88]

Months of work and negotiation had yielded a plan that the President found acceptable. Although seeking greater emphasis on Mars, the President had no major complaints with other aspects of the plan. He expressed support for bringing in international partners and for creating a focused commission to help decide questions regarding implementation, as the proposal recommended.[89] While O'Keefe failed in his last-ditch effort to gain support for a higher budget and closing the gap in the U.S. human spaceflight program, he was satisfied with the outcome. NASA had a new vision, with full support from the White House.

---

86. "Briefing for the President," p. 11; Schumacher notes. Gil Klinger believed that science would be served merely by keeping humans flying in space. He noted in a retrospective interview that "[t]he scientific community was going to whine about this no matter what. If we had picked different destinations, whoever was disenfranchised, would have complained." Somewhat more sensitively, he explained that much of the criticism from the scientific community of human spaceflight has been a product of the government's failure to fund the Earth sciences adequately. Klinger interview, pp. 46–48.

87. Isakowitz indicated that O'Keefe's major concern about dependence on the Russians was that it was "a very dangerous thing to do," not "in terms of human life, but dangerous politically" (Isakowitz interview, pp. 74–75).

88. Schumacher notes; Radzanowski, Record, 22 December 2003, p. 2.

89. Schumacher notes; "Briefing for the President."

A Difference of Opinion over Emphasis

Disagreement about the meaning of the decision arose even before the principals left the room. Marburger recalled that "as we pushed our chairs away from
the table, Sean [O'Keefe] rose to his feet and looked at me and said 'Did you
hear what I heard? We're going to Mars.'" Marburger interpreted the President's
words differently. Aware that Hadley had taken copious notes, Marburger asked
Hadley to type up his notes and circulate them to give all participants in the
meeting an opportunity to concur on what had transpired. Hadley agreed, but
Deputy Chief of Staff Miers objected on the grounds that, as a matter of practice, she did not allow the retention of minutes from meetings advising the
President. Marburger later explained that he could accept a level of ambiguity
in interpretation because he and O'Keefe largely concurred on the details of
the plan even though he viewed O'Keefe's reaction as further proof of NASA's
obsession with a dramatic Mars mission. Marburger viewed the President's
comments at the meeting as supporting a much broader vision, encompassing
robotic and human spaceflight with numerous possible destinations and programs.[90] From the NASA perspective, ironically, Marburger and OSTP had
pushed a narrow vision, focused too heavily on human missions to the Moon.

Regardless, the 19 December meeting was the high point of the policy
development process. It ended all doubt among the leadership of NASA and
the space policy staff at the White House that the space program would move
forward with a new mission—an imperfect mission, but one that had the support of NASA, the President, and key agencies within the executive branch.
Notwithstanding Marburger's concern over how the policy would be "sold"
publicly, the meeting put to rest, at least for the time being, major disputes
over the budget; U.S. access to space, the Moon, and Mars; and the role of the
Shuttle and Space Station in NASA's future plans.

Drafting the NSPD

The meeting with the President was one of many hurdles on the path to implementation of the VSE. The NASA and White House staff that put together
the presentation to the President began working weeks before the meeting on
supporting documents for the formal public announcement of the VSE, as well
as the directive for internal government use only that would establish the VSE
as formal presidential policy. NASA staff and their White House counterparts

---

90. Marburger interview, pp. 34–35.

argued vigorously following the 19 December meeting over emphasis and inter-pretation of the President's decision.

The most vigorous debates occurred in the context of developing National Security Presidential Directive (NSPD)–31, which served as the formal policy document for the VSE. The final draft, signed by the President on 14 January 2004, differed in significant respects from the earliest drafts that had circu-lated among the individuals involved in the interagency working group. With Gil Klinger coordinating all aspects of the document revision process, Steve Isakowitz and John Schumacher went to great lengths to ensure that the docu-ment both reflected the President's priorities and allowed NASA the widest latitude possible for interpreting and implementing the President's direction.

Dismayed that the initial draft NSPD focused too heavily on human space-flight and the Moon and not enough on robotic science goals, Isakowitz rewrote large sections of the document the week before the meeting with the President. Explaining the justification for his revisions, Isakowitz told Schumacher:

> it's reading very narrow—it reads like it's all about sending humans to the Moon. Again, we need to be clear and vigilant on this point that the VSE is much bigger than that Wash Post sound bite ("Apollo redux")—it's about humans AND robots, with robots leading the way now to multiple destinations like Mars/etc., and—oh by the way—we will start with a visit to the Moon.[91]

To make the NSPD better able to withstand criticism and to more accurately reflect NASA's plans and ambitions, Isakowitz and Schumacher sent back a draft to Klinger that highlighted the advances in knowledge and technology NASA had made through robotic scientific exploration in the past decade and drew the connection between human and robotic exploration more strongly, while at the same time emphasizing that "robotic activities are not just precur-sors" to human exploration. Isakowitz also altered language to ensure that it mentioned multiple destinations for human exploration, rather than solely the Moon. His revisions included the first mention of Mars in the NSPD, in fact.[92]

Klinger and the other individuals involved in developing the NSPD were receptive to the revisions Isakowitz and Schumacher proposed. Dave Radzanowski at OMB agreed that the group needed "to be careful that the vision and all associated documents do not come across as being solely about

---

91. Isakowitz to Schumacher, 12 December 2003, Vision Development emails folder.

92. Isakowitz to Klinger, 12 December 2003, Vision Development emails folder; Schumacher interview, p. 25; Isakowitz interview, pp. 89–90; draft NSPD, 10 December 2003, Vision Development emails folder 2 of 3; Klinger to Radzanowski et al., 15 December 2003, attachment, "Draft NSPD with agency inputs," Vision Development emails folder.

humans." He responded to Isakowitz's email on the revisions agreeing that sending humans to the Moon "is just one component of a much broader vision that includes sending multiple and more sophisticated robotic probes throughout the solar system now and in the future."[93] While reaffirming the importance of robotics, Radzanowski's depiction of the VSE appears to have excluded human spaceflight beyond the Moon. Perhaps Radzanowski's omission was accidental and incidental, but it did foreshadow some of the disputes between NASA and OMB that would continue up to the signing of the NSPD.

The two primary issues NASA and OMB fought over were the retirement date for the Shuttle and the relative prominence of the Moon as a destination. Klinger resolved less significant disputes over completion dates, such as for the first test flight of a new CEV and robotic goals, with generous timetables and vague wording. For the Shuttle, early drafts called for retiring the fleet "as soon as construction of the International Space Station is completed, but not later than the year 2010." Isakowitz and Schumacher tried to remove constraints on Shuttle retirement several times, first through simply removing the wording "not later than the year 2010," and then by changing the wording to "planned in 2010."[94] As they saw it, they were remaining mindful of CAIB criticisms about schedule pressures at NASA and trying every way possible to avoid including language that would force the Agency to try to meet a timetable that might compromise safety.[95]

Radzanowski protested vehemently, however, arguing, "We do not want wiggle room on Shuttle retirement date—implementation of VSE relies heavily on Shuttle retirement and the funds it frees up, as well as avoidance of recertification."[96] Radzanowski's superiors at OMB, Joel Kaplan and Marcus

93. Radzanowski to Isakowitz, 15 December 2003, Vision Development emails folder. Others copied on emails regarding the draft NSPD included Schumacher, Mahone, Joe Wood, and Michael O'Brien at NASA; Jim Marrs from the Vice President's Office; Julia Stahl from the EOP Office of Administration; Eric Helland from CEA; Elizabeth Ahern from the NSC; Allison Boyd in the EOP Office of Policy Development; Paul Shawcross, Jason Rothenberg, and Amy Kaminski at OMB; and Bill Jeffrey and Brett Alexander at OSTP.

94. Klinger to Radzanowski et al., 12 December 2003, attachment, draft NSPD, Vision Development emails folder; Klinger to Radzanowski et al., 15 December 2003, attachment, "Draft NSPD with agency inputs," Vision Development emails folder; Radzanowski to Isakowitz, 17 December 2003, Vision Development emails folder.

95. Radzanowski to Isakowitz, 17 December 2003, Vision Development emails folder; Isakowitz interview, pp. 89–90; and Schumacher interview, pp. 25–26.

96. Radzanowski to Isakowitz, 17 December 2003, Vision Development emails folder. CAIB researcher Dwayne Day believes that the CAIB pushed for Shuttle recertification if it were to fly past 2010 because the CAIB members assumed that if the Shuttle returned to flight in 2004 or 2005 and its missions were focused on building the ISS, the ISS could be completed by 2010. Thus, this date was selected for reasons other than the physical conditions of the orbiters.

Peacock, insisted on keeping 2010 as the retirement year. (As it turned out, the last Shuttle mission was in 2011.) Sean O'Keefe called Kaplan in the evening of 23 December to press upon him the importance of giving NASA more flexibility to deal with the Shuttle retirement. O'Keefe made a convincing case. To the chagrin of Radzanowski and others at OMB, Kaplan agreed to remove all firm commitments to Shuttle retirement. The final NSPD simply called for the retirement of the Shuttle upon the completion of the ISS, "planned for the end of this decade." NASA finally gained the flexibility to move the retirement date if circumstances or priorities changed.[97]

NASA fought with OMB and OSTP on the role of the Moon from the start of the writing of the NSPD and throughout December, and the issue came to a head in early January. Gil Klinger sent out a draft of the NSPD on 31 December, insisting that it was "to be reviewed ONLY for factual accuracy."[98] Bill Jeffrey of OSTP responded the following Monday, 5 January 2004, with a few minor changes in wording that fundamentally altered the role of the Moon in the VSE. Rather than the first of many potential destinations for human activity, Jeffrey's edits suggested that an extended human presence on the Moon would be the necessary foundation, the primary source of resources and infrastructure, for all human exploration.[99] In an internal memo to Schumacher, O'Brien, and Isakowitz, Joe Wood, by then a senior official in NASA's Office of External Relations, wrote, "OSTP's changes are obnoxious, greatly emphasizing lunar resources, extended stay on the Moon, and even adding that going to the Moon will significantly decrease the costs of going to Mars."[100] Fortunately for NASA, Klinger responded to the late substantive edits by telling Jeffrey that it was "[t]ime to knock it off."[101] The final NSPD did not include Jeffrey's edits, but it did reflect the tension between NASA's efforts to develop a policy document that enabled a broad range of opportunities and OSTP's attempts to concentrate

---

Day also reminds readers that the CAIB Report pointed out that Shuttle recertification was only one, albeit the main, element of the Shuttle's Service Life Extension Program, which would also address infrastructure and other issues. See Dwayne A. Day, "The Decision to Retire the Space Shuttle," *Space Review* (18 July 2011), available at *http://www.thespacereview. com/article/1887/1* (accessed 17 June 2013).

97. Isakowitz to Schumacher, 19 December 2003, Vision Development emails folder; Schumacher to Klinger, 24 December 2003, Vision Development emails folder; Radzanowski to Schumacher, 29 December 2003, Vision Development emails folder; NSPD-31.

98. Klinger to Radzanowski et al., 31 December 2003, Vision Development emails folder.

99. Jeffrey to Schumacher, 5 January 2004, attachment, NSPD-31 draft, 31 December 2003, Vision Development emails folder.

100. Wood to Schumacher et al., 6 January 2004, Vision Development emails folder.

101. Klinger to Jeffrey, 7 January 2004, Vision Development emails folder.

the VSE solely on the Moon, specifically setting up a permanent lunar base and investing vigorously in mining the Moon.

A Done Deal

While the NSPD more closely reflected NASA's aspirations, the text of President Bush's VSE announcement on 14 January contained the strong lunar settlement and resource exploitation wording that Bill Jeffrey had promoted, including the concept that "an extended human presence on the Moon could vastly reduce the costs of further space exploration."[102] Isakowitz, for one, was surprised at the divergence between the NSPD and the speech, particularly the limited mention of Mars and the strong emphasis on the Moon in the President's speech.[103]

Yet NASA did not fret any further about the subtle details of the announcement or the President's intentions. Isakowitz, Schumacher, and O'Keefe had accomplished what they set out to do in the context of the interagency negotiation process. They did not get the VSE they wanted in every respect. The White House denied NASA's initial ambitions for a more expensive and elaborate vision that would have maintained U.S. independence in space. NASA, however, also avoided being locked into a rigid and costly program with ambitious deadlines that would have absorbed all of the Agency's future budget increases and produced little of value for the public, the spaceflight engineering community, and the space science community.

The vision that the interagency process produced ultimately gave NASA a great deal of flexibility on destinations, timetables, and goals. Beyond providing NASA with a path to move beyond the Space Shuttle, Gil Klinger, the person who coordinated the policy development process, had few expectations for the VSE. He believed that the White House should work with NASA leaders to give the Agency "broad strategic direction" but that the details of the program and the difficult decisions about tradeoffs between programs should be left to NASA. Klinger also wanted to avoid imposing rigid goals or timelines that might swell the Agency's budget or undermine the entire program.[104]

NASA saw the process in a similar light. Although the VSE emphasized the Moon and forced the Agency to retire the Shuttle before a new CEV became available, Isakowitz felt that he and his colleagues had succeeded in keeping the NSPD free of restrictive language about the Moon and other priorities. In agreeing to the final language of the NSPD, the NASA team ultimately decided

---

102. Bush, Vision speech.
103. Isakowitz interview, pp. 89–90.
104. Klinger interview, pp. 36–41.

that meeting the VSE's obligation to reach the Moon would not be difficult. The nature and extent of Moon exploration, as they saw it, remained a decision for NASA.[105] Nothing in the NSPD prevented the Agency from changing course on elements of the Moon program and other details beyond the Moon should priorities, circumstances, or goals change. In this respect, the VSE enabled NASA to choose its own path in expanding human and robotic exploration throughout the solar system.

---

105. Isakowitz interview, pp. 88–89.

7

IMPLEMENTING THE VISION FOR SPACE EXPLORATION

THE PRESIDENT'S 14 January 2004 Vision for Space Exploration speech[1] provoked mixed reactions at home and abroad. While many heralded the announcement at NASA Headquarters as the next great catalyst for America's space program, others expressed caution and questioned the plan's utility and feasibility, as well as the motives behind it. House Majority Leader Tom DeLay (R-TX) justified support for the VSE with sentimental words that were at once personal and political: "NASA helps America fulfill the dreams of the heart."[2] Many VSE supporters shared Delay's nationalistic sentiments, asserting that the desire to explore and take on new challenges was the essence of what it means to be American. To these supporters, history and fate placed a special burden on the United States to adopt this new policy for the benefit of humanity.

A number of prominent supporters of the space program expressed concern that the United States lacked a strong rationale or motivation to justify sustained support for an ambitious space exploration plan. John Logsdon, who was then the Director of the Space Policy Institute at George Washington University, argued that "space exploration has never been a powerful motivator for those controlling the resources to make it happen."[3] Logsdon and others worried that without the kind of political conditions that had allowed for continued support

---

1. Bush, Vision speech.
2. Guy Gugliotta, "DeLay's Push Helps Deliver NASA Funds," *Washington Post* (6 December 2004). DeLay's 22nd District of Texas initially abutted and later included Johnson Space Center, so he had a local political incentive to support human spaceflight.
3. John Logsdon, "Which Direction in Space?" *Space News* (4 October 2004): 13.

of the Apollo program during the Cold War, future administrations would lack the incentive to follow through with the VSE's implementation.

Economic opportunity gave companies and communities already connected to NASA a compelling reason to support the VSE. Companies such as Lockheed Martin, Boeing, and Ball Aerospace and Technologies Corporation prepared to fight for their share of the business generated by the VSE, as they believed that the new technologies and infrastructure necessary for implementing the space exploration policy would generate lucrative government contracts and create jobs.[4] In the days immediately following the VSE announcement, political officials and corporate representatives spoke out to draw attention to the possible contributions of their communities and companies to the VSE. Louisiana Governor Kathleen Blanco spoke with President Bush about the role that NASA's Michoud Facility in New Orleans could play in the VSE's implementation.[5] Corporate spokesman Marion LaNasa cited Lockheed Martin's unique composite materials production and metal works capabilities in the hope that NASA would realize the utility of maintaining a strong partnership with the company.[6]

NASA encouraged competition between the various companies jockeying for roles in the VSE by selecting 11 teams for six-month study contracts worth a total of $27 million. Eight teams won $3 million each to study Crew Exploration Vehicle concepts as well as lunar exploration architectures. Three other teams were awarded approximately $1 million each solely to examine lunar exploration options. NASA selected large, established companies such as Boeing and Lockheed Martin, which already held major NASA engineering contracts, as well as lesser-known teams from Andrews Space, Inc.; Charles Stark Draper Laboratory; and Schafer Corporation. By not allowing these teams to share CEV designs, NASA hoped to create a competitive environment that drew forth multiple independent designs and ultimately generated a final product of the highest quality.[7]

Aerospace companies nonetheless perceived their common interests in promoting the VSE. Boeing and Lockheed Martin combined to form the nucleus of the Coalition for Space Exploration, which promoted the VSE to the public, lawmakers, and other space exploration stakeholders. Lockheed Martin

---

4. Gregg Fields, "President's Moonwalk Proposal Gets Mixed Reaction," *Miami Herald* (15 January 2004).

5. Associated Press Newswires, "Officials to Lobby for Role in Bush's Space Plan," 16 January 2004.

6. Ibid.

7. Brian Berger, "NASA Diversifies Awards of Lunar Exploration Contracts," *Space News* (6 September 2004): 3.

Vice President John Karas noted at the coalition's inception that the aerospace industry's job is not merely to provide technical expertise. Coalition members believed that the ability to increase public awareness was vital to the success and advancement of the projects on which the industry as a whole depended.[8]

International Reception

During his VSE announcement, President Bush emphasized that his plan "is a journey, not a race," and asked America's international partners to support the new space exploration policy.[9] In more formal language, the VSE called for NASA to "pursue opportunities for international participation to support U.S. space exploration goals."[10] Besides honoring national commitments to completing the International Space Station, it was not clear how much the administration *wanted* international collaboration on the new aspects of the VSE (going back to the Moon and on to Mars with robots and humans). Thus, not surprisingly, the President's new space policy received mixed reactions from abroad.

Despite President Bush's assurances that the U.S. would honor its commitment to complete its work on the ISS by 2010, international partners focused on other aspects of the policy that would immediately impact their investments in the ISS, such as the Vision's downgrading the priority of ISS international modules and labs.[11] Many international partners questioned America's willingness to maintain its role on the ISS given the enormous projected costs for the VSE. The ISS was "ESA's near-term focus for exploration" and thus ESA viewed the "Exploration Vision in the context of ongoing ISS activities."[12]

8. Brian Berger, "Rival Firms Unite Behind U.S. Space Exploration Plan," *Space News* (29 March 2004): 40.

9. Bush, Vision speech.

10. *The Vision for Space Exploration* (Washington, DC: NASA, February 2004), p. vi.

11. White House Office of the Press Secretary, "President Bush Announces New Vision for Space Exploration Program, Fact Sheet: A Renewed Spirit of Discovery," 14 January 2004, available at *https://history.nasa.gov/SEP%20Press%20Release.htm* (accessed 4 May 2018). In the "Vision for Space Exploration Fact Sheet," February 2004, p. iv (p. 6 overall in the PDF), available at *https://history.nasa.gov/Vision_For_Space_Exploration.pdf* (accessed 4 May 2018), the ISS component of the VSE is focused on "supporting space exploration goals, with emphasis on understanding how the space environment affects astronaut health." See also John Krige, Angelina Long Callahan, and Ashok Maharaj, *NASA in the World: Fifty Years of International Collaboration in Space* (New York: Palgrave Macmillan, 2013), p. 194. Krige et al. also cite Carl E. Behrens, *The International Space Station and the Space Shuttle* (Washington, DC: Congressional Research Service Report 7-5700, 18 March 2009).

12. Stephen Ballard, International Programs Specialist, Memo for the Record, "Meeting Between Mr. Marco Caporicci…European Space Agency; and Dr. John Rogacki, Office of Aeronautics, March 4, 2004," 10 March 2004, p. 1, OER-internationalfiles, NASA HRC.

German space officials were particularly frustrated. They had been stung in spring 2002 by NASA's seemingly sudden cancellation of the X-38—a project to develop a space plane that would act as an emergency "lifeboat" for ISS crew—which involved substantial contributions from the European Space Agency (ESA) collectively and the *Deutsches Zentrum für Luft- und Raumfahrt e.V.*, or German Space Agency (DLR), in particular.[13] In meetings with NASA officials two years later, DLR leaders raised the X-38 cancellation as their first topic and emphasized that there could be "no underestimating the negative pall cancellation of X-38 casts over future U.S.-German cooperation." The lack of ISS coordination in the VSE added insult to injury in German minds. DLR officials had made substantial investments in the ISS and felt that the United States was now walking away from its commitment. Additionally, DLR leaders "share[d] the general skepticism evidenced by many European space officials on the U.S.'s ability to sustain The Vision over a long duration."[14]

Like the Germans, the Danish were critical of the lack of defined opportunities for international partners to participate in NASA's new Vision and doubted the ability of the United States to adequately fund and sustain it. Officials at the Danish Space Research Institute (DSRI), with their successful record of cooperation with NASA on several robotic Mars experiments in mind, "proposed a more highly integrated, truly international approach to cooperation, not cooperation on the U.S.-defined vision and goals."[15]

European partners, separately, as well as collectively in such entities as ESA and the European Union, were disappointed with their lack of opportunity to provide input on the Vision and were concerned with the impact it would have on their own priorities for space. Almost all had cooperated on robotic space science projects and wanted greater emphasis placed on robotic missions.[16] ESA, furthermore, had already begun initial studies for a human-robotic program for Moon-Mars exploration called Aurora.[17]

---

13. See Mark Carreau, "X-38 Project's Cancellation Irks NASA, Partners," *Houston Chronicle* (9 June 2002), available at *http://www.chron.com/news/nation-world/article/X-38-project-s-cancellation-irks-NASA-partners-2064969.php* (accessed 10 January 2018).

14. J. Donald Miller, NASA European Representative, Memo for the Record, "NASA Meetings in German[y] and Denmark," 17–18 May 2004, in OER-internationalfiles/Tawney-Europedocs/ folder, NASA HRC; see p. 2 for both quotes.

15. Ibid., p. 3.

16. See, for example, Stephen Ballard, International Programs Specialist, Memo for the Record, "Meeting Between Mr. Marco Caporicci…European Space Agency; and Dr. John Rogacki, Office of Aeronautics, March 4, 2004," 10 March 2004, p. 3, OER-internationalfiles, NASA HRC.

17. See "The Aurora Programme: Europe's Framework for Space Exploration," *ESA Bulletin* 126, May 2006, *http://www.esa.int/esapub/bulletin/bulletin126/bul126b_messina.pdf* (accessed

Canadian Space Agency (CSA) officials expressed misgivings about the VSE's vagueness—specifically, its lack of a formal beginning or end—and asked how a new "vision" in 10 years could be avoided. CSA officials pressed NASA colleagues on congressional views and whether NASA genuinely wanted international input on the VSE. When a CSA leader asked for a presidential letter to leaders of potential partner nations to shore up support for the VSE, a NASA official replied that the President's January speech was the de facto invitation to participate.[18]

Russian space officials remained cool toward President Bush's VSE announcement. They initially claimed that they could implement a similar policy at a fraction of the cost. Nikolai Moiseyev, deputy director of the Russian Federal Space Agency (also known as Roscosmos), asserted that Russia possessed the technological capabilities to begin launching its own human lunar and Martian exploration program by 2015 at a cost of $15 billion (1/10 the cost of the American plan).[19] In 2004, Russia also moved forward with, at least on paper, a multipurpose orbital space plane called Kliper that seemed to be a potential competitor to what became NASA's Crew Exploration Vehicle.[20]

At virtually the same time, however, Roscosmos head Anatoly Perminov expressed a desire for his agency to cooperate with NASA since a human spaceflight program was becoming prohibitively expensive for any single nation to operate.[21] Russian aerospace officials also complained that a congressional decision to cut funding for space exploration put an excessive financial burden on Russia. Perminov protested, "against its will, Russia has once again found itself involved in a costly high-tech project with vague prospects."[22]

In immediate reaction to President Bush's announcement, Vyacheslav Mikhailichenko, a Roscosmos spokesman, stated that "they are plans—they

25 January 2018); "Asia, Europe Give Nuanced Welcome to Bush's Space Dream," *Agence France-Presse* (15 January 2004); and "European Space Agency Gives Cautious Welcome to Bush Plans for Space Travel," *Agence France-Presse* (16 January 2004).

18. Timothy Tawney, Memo for the Record, "Deputy Assistant Administrator Visit to Canadian Space Agency Headquarters to Discuss the U.S. Vision for Space Exploration," 17 June 2004, pp. 2–3.

19. "After Bush Speech, Russia Mulls Missions to Moon and Even Mars," *Agence France-Presse* (15 January 2004).

20. See, for example, "Russia Building New Space Shuttle," *RIA Novosty* (19 April 2004); "Russian TV Shows New Space Shuttle Kliper," *BBC Monitoring Former Soviet Union* (13 May 2004); and "Space Agency Suggests Making Kliper Project International," *Interfax News Service* (28 June 2004).

21. "Russia Wants To Participate in U.S. Space Program," *Prime-TASS News Agency* (31 March 2004).

22. "U.S. Congress Decision To Cut Space Funds Seen as Placing Heavy Burden on Russia," *Moscow Nezavisimaya Gazeta* (2 August 2004): 1, 7.

don't affect us in any way" and that Russia would not abandon its commitment to the ISS.[23] The first part of this statement likely indicated a nonchalant public pose, in the sense that many proposed policies are never seen through to fruition. Any public rhetoric implying that the Russians could finish the ISS themselves, however, was disingenuous since the technical and financial designs of this project were inherently *international*. Reinforcing this tenet from a different perspective, key Russian officials asserted that no single country could undertake Mars exploration successfully—implying that Russia was an essential partner. Other space agencies quickly adopted this mantra, including NASA.[24]

Later that year, an unnamed Russian journalist opined that the "ISS, with its construction currently half completed, will never exist in the originally planned form…. [Russia] will carry the main burden of keeping the ISS in the minimum required operational condition…for at least 15 years."[25] This journalist also argued that "NASA has given up on further ISS development…[i]n full compliance with the Russian proverb 'Friends are OK when they don't get in the way.'"[26] Roscosmos announced that summer that it would "stop being a free-of-charge 'space cabbie' for NASA." Perminov was quoted as saying that "if in 2005 the Americans wish to fly on Soyuz craft, then let them pay for the costs of the flight" given that "the construction of the ISS had stopped, putting the whole burden of rotating crews and delivering cargo to orbit on Russia."[27]

Leaders in Japan were reluctant to commit to a plan they considered vague and undefined. The Japanese wanted to see more details before making a decision to support the VSE.[28] Japan, having already suffered from ISS cost overruns and a tight space budget, stood to lose much of its investment if the United States diverted resources away from the program. Still, in July 2004, a Japanese space official told a NASA colleague that he was envious that NASA had received positive presidential attention, whereas the Japan Aerospace Exploration

---

23. Sonia Oxley, "Russians Cool, but Europe Warms to U.S. Space Plan," *Reuters News* (15 January 2004).

24. "Russia Wants To Participate in U.S. Space Program," *Prime-TASS News Agency* (31 March 2004). This article paraphrases Perminov as saying, "Within the next five years, no country, including the richest ones, will be able to organize an expedition to Mars on its own." See also "Summary of Discussion, Eighth Meeting of the U.S./Russia Space Science Executive Joint Working Group," Babakin Science Center, Moscow, 8 June 2004, p. 2 (electronic copy in OER-internationalfiles/Barry-Russianfiles, NASA HRC).

25. "U.S. Congress Decision To Cut Space Funds Seen as Placing Heavy Burden on Russia," *Moscow Nezavisimaya Gazeta* (2 August 2004): 1, 7.

26. "U.S. Congress Decision to Cut Space Funds Seen as Placing Heavy Burden on Russia," pp. 1, 7.

27. "Russia No Longer Wishes To Be 'Free-of-Charge Space Cabbie' for US," *ITAR-TASS* (4 August 2004). All three quotes from the above two sentences are from this source.

28. "Asia, Europe Give Nuanced Welcome to Bush's Space Dream."

Agency (JAXA) was struggling to develop a long-term space strategy going beyond their five-year plan.[29]

Simply put, the main political issue regarding the VSE was that many international partners felt blindsided by what they viewed as the Bush administration's unilateral decision to change course on the technologically complex and diplomatically challenging ISS project. After the policy was unveiled, administration officials had few good options for assuaging the concerns of partners. While the architects of the VSE may have had good reasons for not reaching out earlier to partners, including the desire to come to agreement domestically before raising expectations internationally, even a modest effort to inform and gain support from international space partners would have helped the Agency avoid the tensions it encountered.

The President, nevertheless, avoided the strongest possible resistance from both allies and adversaries by downplaying the VSE's potential military impact and insisting that his new plan would not provoke a Cold War–style race. China's successful launch of a taikonaut into orbit aboard a Shenzhou 5 spacecraft in October 2003 suggested to some that a threat to U.S. national security interests could emerge if the Chinese government continued to develop sophisticated technology.[30] The China National Space Administration (CNSA) echoed President Bush's assertion that space exploration should not draw the two countries into a hostile competition.[31] Russian spokesman Mikhailichenko also denied that the VSE would lead to a renewed space race, as the Russian space agency did not have the resources to maintain a competition and was content to continue with its own research and development plans.[32] Supporters in the United States pointed to President Bush's comment about the VSE being a "journey, not a race" to deny the possibility of a renewed space race or military element to the VSE. Although the VSE was neither framed nor interpreted as an effort to gain military advantage over potential adversaries in space, the desire of the United States to maintain international leadership in space loomed large. For a nation facing an array of new competitive and cooperative

---

29. Chris Blackerby and Gib Kirkham, NASA International Program Specialists, Memo for the Record, "Admiral Steidle's International Meetings at the Farnborough International Air Show," Farnborough, U.K., 18–21 July 2004, , "OER-internationalfiles/Tawney-Europedocs/Scan1.pdf," DPT/VSE files, NASA HRC.

30. Gwyneth K. Shaw, "Bush's Space Plan Draws Raves," *Orlando Sentinel* (15 January 2004).

31. Kong Quan, a Chinese Foreign Ministry spokesman, for example, blandly stated that "China, in the field of spaceflight and exploration of space, hopes to cooperate with other countries on the basis of equality and friendship." See "UPDATE 1-Both China and the United States Set Sights on the Moon," *Reuters News* (15 January 2004).

32. Sonia Oxley, "Russians Cool, but Europe Warms to U.S. Space Plan," *Reuters News* (15 January 2004).

relationships in space, the VSE promised to reinvigorate the idea of the United States as a technological powerhouse and thereby give its leaders and citizens a renewed sense of importance in the international community.

Congressional Reaction

Even though many lawmakers questioned President Bush's motivations and willingness to follow through with his initiative, the VSE won favor among many members of Congress. The Republican Party offered enthusiastic support for the announcement, which came at the beginning of an election year. House Majority Leader Tom DeLay declared that "naysayers and short-sighted people…cannot stop the American spirit," fueled by the "juggernaut of Americans' excitement about exploring the unknown." Representative Tom Feeney (R-FL) supported DeLay's claim, saying that "the President's proposal is bold, it's exciting, it's visionary, [and] it's affordable." Congressman Sherwood Boehlert (R-NY) offered a more sober take on the VSE, asserting that the VSE "is something that's doable, but it's going to be a challenge."[33] Focusing on the political hurdles to the VSE's implementation, Boehlert hinted that lawmakers likely would vigorously debate the plan.

One significant criticism of the VSE was that the President did not provide sufficient funding to sustain it, even in the near term. President Bush initially proposed adding $1 billion, spread over five years, to NASA's budget. Given NASA's annual budget of approximately $16 billion at that time, the additional funding amounted to an increase of approximately 1 percent each year. The President called for the remaining costs to be absorbed by NASA's human spaceflight effort without causing any significant adverse impact to other NASA programs, such as space science, Earth science, or aeronautics.[34] While Bush's FY 2005 budget proposal (submitted in February 2004) called for a 4.3 percent increase in research and development spending, most of this amount was for military and homeland security programs.[35] NASA's actual FY 2005 budget authorization of $16.198 billion was approximately 5 percent more than the total of $15.379 billion for FY 2004. Yet, adjusted for inflation,

---

33. All three quotations are from Shaw, "Bush's Space Plan Draws Raves."

34. In addition to the *http://history.nasa.gov/Bush%20SEP.htm* site (accessed 7 May 2018), see *http://history.nasa.gov/sepbudgetchart.pdf* (accessed 7 May 2018) for a copy of a well-publicized, but little-understood, chart depicting this new exploration program's budget.

35. "From the Hill: Defense, Homeland Security Dominate Bush's FY 2005 R&D Budget," *Issues in Science and Technology* (spring 2004): 17; Jim Dawson and Paul Guinnessy, "Bush R&D Budget Remains Focused on War, Terrorism, and Security in FY 2005; Civilian R&D Funding Flat," *Physics Today* 57, no. 4 (1 April 2004): 35–41.

the percentage increase was about 2.8 percent—a real increase, and more than what Bush initially proposed, but likely insufficient for maintaining spending on existing commitments and initiating new projects to support a major space exploration initiative.[36]

The President and his critics needed to look no further than across the Potomac, to the Pentagon, for a sense of what a major new aerospace technology initiative might cost. Work on the Air Force's F-22A Raptor, a stealth fighter jet that involved groundbreaking research and sophisticated engineering, had begun in 1986. Development of the system had begun in 1991. By the time the system achieved initial operational capability in 2005, the Air Force had spent $46 billion, an average of over $3 billion per year, on research, development, and testing. Projections for the total cost of the Raptor decreased from $81 billion in 1992 to $65 billion in 2004 as the total anticipated buy for the aircraft plummeted from 648 to just 181. Cost per unit had grown over the same period by almost threefold, from $125 million in FY 1992 to $361 million in FY 2004.[37] While achieving a stable and reliable design to produce the F-22A Raptor in quantity posed major challenges, Raptor engineers did not venture to the cutting edge of aircraft technology on their own. The system was one in a long line of sophisticated fighter jets with highly complex software and electronics. Raptor advances in software, avionics, aerodynamics, and stealth technology were a matter of degree; significant, but they still represented only incremental advances over past fighter jets. Producing the systems needed to support the VSE, on the other hand, would require venturing into technological domains in which aerospace engineers had scant experience, particularly since VSE systems would follow a very different technological path from that of the Space Shuttle or other existing space systems. Thus, the F-22A example raises the question of whether the budget Congress provided for the VSE would have been sufficient for doing anything more than establishing a skeletal organization to run the program, initiating the proposal development process, and redirecting existing personnel and programs to begin preliminary research on a new launch system. A serious innovation program to support the VSE, involving research,

36. Figures derived from Appendices D-1A and D-1B of the *Aeronautics and Space Report of the President, Fiscal Year 2008 Activities* (Washington, DC: NASA NP-2009-05-581-HQ, 2009), available at *http://history.nasa.gov/presrep2008.pdf* (accessed 7 May 2018), and annual inflation rates available at *http://www.usinflationcalculator.com/inflation/current-inflation-rates/* (accessed 8 September 2011).

37. *Aeronautics and Space Report of the President, Fiscal Year 2005 Activities* (Washington, DC: NASA NP-2007-03-465-HQ, 2007), available at *http://history.nasa.gov/presrep2005.pdf* (accessed 7 May 2018); Government Accountability Office, *Defense Acquisitions: Assessments of Selected Major Weapon Programs* (Washington, DC: GAO-06391, March 2006), pp. 59, 71.

development, testing, and production, would take billions of dollars over many years. An immediate injection of a couple hundred million dollars would be a weak start if not followed shortly thereafter with dramatic and sustained increases in NASA's exploration budget.

Whatever the ultimate cost of the program, some critics felt that the money dedicated to the VSE would be better spent on more immediate, Earthbound problems, such as education, health care, terrorism, and the Iraq war.[38] Senator Ernest Hollings (D-SC) summed up the attitude of many lawmakers—indeed, many in the United States—when he stated that the Bush administration's "interest doesn't reflect an honest assessment of the fiscal and organizational realities facing NASA and the financial realities facing the country."[39] Many even saw the announcement as an election stunt designed to increase the President's public approval rating and assist his reelection campaign in much the same way that the public rallied around the Apollo program in the 1960s.[40] Still others recognized that the VSE remained undefined, with major details left to be revealed or worked out at a later date. Senator Bill Nelson (D-FL), though skeptical, simply allowed that he wanted to "see what the details are." One critical Senate staffer complained that the VSE was "nothing but vagueness."[41]

Not all observers of the space program felt that the VSE's vague language would hinder its ultimate success. John Logsdon suggested that the President should provide general guidelines for the VSE and allow NASA to define it in time. "NASA will have to come forward with the details as they present their budget and go through the congressional hearing process," he said. "There's an impatience to learn everything in one day, but I think that realistically, we have to be patient."[42]

The scientific community feared that the VSE's emphasis on human space exploration would drain resources from programs that did not involve human spaceflight, including those in the Earth and planetary sciences that had been integral to NASA's mission in the past. Robert Kirshner, an astronomy professor at Harvard University, echoed the sentiments of many scientists when he

---

38. "Americans Welcome Space Goals, but Skepticism over Cost," *Agence France-Presse* (15 January 2004).

39. "Americans Welcome Space Goals."

40. Mark Henderson, "Great Election Ploy, but Where Will Bush Find the Billions?" *The Times* (10 January 2004). The CAIB Report also noted the "lack, over the past three decades, of a national mandate providing NASA a compelling mission requiring human presence in space." (See p. 209 of the CAIB Report.) For more on the context after the Columbia accident, see chapter 5.

41. Both quotations are from Shaw, "Bush's Space Plan Draws Raves."

42. Shaw, "Bush's Space Plan Draws Raves."

complained that "they are doing [the VSE] from the top down."[43] However, whether the Vision's science goals were added after the fact or developed as an outgrowth of DPT/NEXT planning (or some combination of both) is an open question. Scientists who worked closely with NASA and depended on the Agency for support felt that they should have been consulted to ensure that the VSE reflected the goals of the space science community. To these scientists, the VSE represented a political stunt, relying on grand goals and promises of achievements in human spaceflight to bolster public confidence in the administration. That cost overruns for the Space Shuttle and ISS threatened funding for science programs—especially given that the NASA Administrator could shift funding from other programs to the ISS and Shuttle—inspired further angst among NASA scientists and NASA-funded scientists who already felt that their programs did not receive appropriate funding.[44]

The Aldridge Commission

In his 14 January 2004 speech unveiling the VSE, President Bush announced that he was setting up a blue-ribbon commission to offer NASA recommendations on how to implement the policy. The President's Commission on the Implementation of United States Space Exploration Policy was known informally as the Aldridge Commission after its chairman, Edward C. Aldridge, Jr., whose career in national security leadership positions included terms as the Director of the National Reconnaissance Office (1981–1988); Secretary of the Air Force (1986–1988); and Under Secretary of Defense for Acquisition, Technology, and Logistics (2001–2003). Other members of the Commission included lunar exploration advocate Paul Spudis, astronomer Neil de Grasse Tyson, and former Hewlett-Packard executive Carly Fiorina. The Commission, also referred to loosely as the "Moon, Mars, and beyond" panel, actively solicited public input through five televised hearings across the country, heard testimony from almost 100 experts, and received thousands of written inputs through its public Web site. Public support for the VSE policy was strong, with

---

43. Cornelia Dean, "At NASA, Clouds Are What You Zoom Through To Get to Mars," *New York Times* (21 March 2005). Presumably, Kirshner meant to imply that the Vision's science goals were imposed from above with little or no consultation with the actual space science community of researchers. The DPT/NEXT science goals were developed in accordance with the National Academies of Science/National Research Council's decadal plans, as well as the former Office of Space Science's strategic plan. For more on this, see Harley Thronson's 25 March 2005 personal email to Robert Kirshner, NASA HRC.
44. Dean, "At NASA, Clouds Are What You Zoom Through To Get to Mars."

approximately 7:1 comments running in favor. The panel publicly submitted its report in June 2004.[45]

The panel recommended broad organizational changes as well as changes in NASA's approach to developing new systems and working with industry. To support the highest levels of decision-making for the VSE, the Commission suggested a "Space Exploration Steering Council," akin to the old National Space Council, that would be chaired by the Vice President and report to the President. Suggestions for improving oversight and management included the creation of several new organizational bodies, including a technical advisory board, an independent cost-estimating division, and an advanced technology division.[46] The report also recommended converting all of NASA's Field Centers into quasi-Government Federally Funded Research and Development Centers (FFRDCs). (JPL was the only NASA Center that was an FFRDC in 2004.)

Reflecting a sober, realistic understanding of the budget President Bush proposed for the VSE, the Commission asserted that the Agency would need to adopt a go-as-you-pay approach, allowing funding levels to determine the pace of the program. Endorsing more commercialization of space, the panel members proposed that private industry take the lead in providing transportation to LEO. The panel also made strong recommendations for engaging international partners and using the Vision to inspire students and further science, technology, engineering, and mathematics (STEM) education. In terms of management, the panel recommended a systems engineering approach that had worked well during the Apollo program, as well as the "spiral development" approach then popular in the military acquisition community. Overall, the report strongly endorsed the President's plan.[47]

The Exploration Systems Mission Directorate Starts Up

The day after President Bush unveiled the VSE, NASA announced a major restructuring of its organization. Craig E. Steidle, who had been serving as a consultant to NASA, was appointed Associate Administrator for a new

---

45. President's Commission on Implementation of United States Space Exploration Policy, *A Journey to Inspire, Innovate, and Discover* (Washington, DC: 2004.) The report is available at *http://history.nasa.gov/aldridge_commission_report_june2004.pdf* (accessed 7 May 2018). For information on public input into the panel's work, see p. 2.

46. NASA has had various technical advisory groups over the years, including the Chief Engineer's Office, formal NASA advisory committees, and an independent NASA safety panel located at NASA Langley Research Center. NASA started a Program Analysis and Evaluation Office in April 2005. NASA had a Chief Technology Officer in the 1990s and then again under Administrator Charles F. Bolden in 2009.

47. Aldridge Commission, *A Journey to Inspire, Innovate and Discover*, pp. 6–9.

Office of Exploration Systems.[48] Cobbled together from components of the aeronautics, spaceflight, and space sciences programs, the Office of Exploration Systems merged with NASA's Office of Biological and Physical Research and officially became the Exploration Systems Mission Directorate (ESMD) on 1 August 2004. The merger and renaming were part of a larger reorganization by Sean O'Keefe to accommodate the VSE and simplify the management of NASA Headquarters. Steidle and his team spent much of the first year expanding the organization with new hires, defining requirements for key elements of the program, and soliciting ideas from industry and the space

Admiral Craig Steidle, the Associate Administrator for Exploration Systems. (NASA Steidle_20040205)

engineering community. In March 2005, ESMD reached a milestone when it released its first Request for Proposals (RFP) for the Crew Exploration Vehicle, which was among the key technologies needed to carry out the program, by then known as Constellation.[49]

A retired Navy rear admiral who was well known for leading major Pentagon acquisition efforts, such as the Navy's F/A-18 program and the Joint Strike Fighter program, Steidle planned to apply the engineering management approach known as spiral development to the Constellation Program. First used in the software industry, spiral development established a routine for incorporating upgrades to existing software programs. Rather than create an entirely new software program each time new capabilities emerged, software developers learned to design systems to allow for upgrades while disturbing existing programs and users as little as possible. Applied to space technology, spiral development would allow new technologies to be incorporated into existing systems as the design matured. Theoretically, the systems would be operational sooner, and upgrades would be less expensive than with previous space systems because the initial design would facilitate upgrading.

---

48. See Glenn Mahone and Bob Jacobs, "NASA Announces New Headquarters Management Alignment," NASA News Release 04-024, 15 January 2004, available at *http://www.nasa.gov/home/hqnews/2004/jan/HQ_04024_alignment.html* (accessed 15 May 2008).

49. "Exploration Systems: A Living History," version 1.0, 19 May 2005, available in NASA HRC.

Steidle proposed a set of spirals that set broad benchmarks such as building the new spacecraft technologies necessary for human spaceflight in LEO, then the technologies necessary for human lunar flights, and taking other intermediate steps before progressing to Mars and deep space operations. The philosophy behind spiral development was to preserve as much flexibility and "evolvability" as possible at each stage of spacecraft development. By making a complex system operational while still tweaking its capabilities, engineers would have the advantages of accelerated development without the difficult retrofitting that often occurs when attempting to incorporate new capabilities or when elements of a system are not fully compatible.[50] Steidle articulated his point of view in a December 2004 interim strategy document:

> This process of flowing from our strategy to program tasks is iterative. Like our overall efforts, the strategy-to-task process is spiral in nature in that, through repeated analysis of costs, performance options, trends, and results—including progress in developing specific capabilities and progress in maturing essential technologies—we spiral towards the deployment of new, transformational capabilities in a manner that is effective and affordable.[51]

While spiral development appeared to be compatible with the go-as-you-pay approach advocated by the Aldridge Commission (and earlier by the Augustine Commission and the Decadal Planning Team), it carried with it disadvantages that its advocates either did not understand or were hesitant to recognize. Critics have argued that the approach gives developers too much flexibility to avoid oversight and ignore cost overruns. Rather than spiraling toward successively advanced iterations of new systems, "spiraling out of control" is the more likely outcome of spiral development, according to one article on the topic.[52] Spiral development had a short life at NASA, but it fit well for approximately a year after the VSE announcement, a period characterized by constrained budgets and optimistic projections for achieving the VSE.

---

50. See, for example, Frank Sietzen, Jr., "Heir to Apollo: Shaping the CEV, Part I: Managing an Exploration Roadmap," *Aerospace America* (January 2006): 31–32; Leonard David, "A Spiral Stairway to the Moon and Beyond," Space.com, 9 February 2005, available at *http://www.space.com/businesstechnology/technology/next_cev_050209.html* (accessed 7 May 2018); and Frank Sietzen, "A New 'Constellation' at NASA," *UPI* (11 May 2004), available at *http://www.spacedaily.com/news/spacetravel-04y.html* (accessed 27 October 2011).

51. The quotation is from the cover letter in the NASA Exploration Systems Interim Strategy, 2 December 2004, available at *http://www.hq.nasa.gov/office/hqlibrary/documents/o56554650.pdf* (accessed 15 May 2008).

52. Victoria Samson, "Spiraling Out of Control: How Missile Defense's Acquisition Strategy Is Setting a Dangerous Precedent," *Defense & Security Analysis* 24, no. 2 (1 June 2003): 203–211.

O'Keefe Departs; Griffin Arrives

Administrator O'Keefe left NASA a little more than a year after President Bush announced the Vision. In February 2005, after having served more than three years filled with distinct highs and one major accident, O'Keefe departed Washington to become the chancellor of the Louisiana State University system.[53] Fred Gregory served as acting NASA Administrator for about two months until a permanent successor was named.

On 14 April 2005, Michael Griffin joined NASA as the new Administrator. An aerospace engineer with multiple advanced degrees, Griffin had a wealth of experience working on civilian and military space programs. Immediately before becoming NASA Administrator, Griffin had served as head of the Space Department at Johns Hopkins University's Applied Physics Laboratory. Earlier in his career, he had served as NASA's Chief Engineer and Associate Administrator for Exploration, working to implement President George H. W. Bush's Space Exploration Initiative. At the Pentagon, he had served as Deputy for Technology at the Strategic Defense Initiative Office. He also had experience working with cutting-edge technologies for the intelligence community, serving as president and chief operating officer of In-Q-Tel, the innovative venture capital arm of the Central Intelligence Agency.[54] In many ways, Griffin was the polar opposite of his immediate predecessor. While he lacked O'Keefe's political skills, which

NASA Administrator Michael Griffin. (NASA)

53. See, for example, Brian Berger, "Three Years in the Hot Seat," *Space News* (20 December 2004); and Guy Gugliotta, "NASA Chief Formally Steps Down," *Washington Post* (14 December 2004). After several years in Louisiana, O'Keefe returned to Washington and the aerospace sector, first as head of General Electric (GE) Aviation's Washington operations and then as Chief Executive Officer of the European Aeronautic Defence and Space Company (EADS) North America. See Brian Berger, "GE Job Marks Return to Aerospace for O'Keefe," *Space News* (7 April 2008); and (anonymous), "Former NASA Chief O'Keefe To Run EADS North America," *Space News* (26 October 2009).

54. Griffin has a Ph.D. in aerospace engineering, as well as master's degrees in electrical engineering, applied physics, business administration, civil engineering, and aerospace science. See *http://www.nasa.gov/about/highlights/griffin_bio.html* (accessed 10 August 2011) for a brief biographical sketch including his educational background and career highlights.

were vital for negotiating the Vision policy, Griffin had a highly unusual combination of depth and breadth of technical expertise, which gave him an ability to independently assess complex technical matters and great confidence in deciding the best technical approaches for implementing the new policy.

After the announcement of the Vision but before he became the NASA Administrator, Griffin jointly led a study commissioned by the Planetary Society to assess the cost and technical feasibility of sending humans to Mars. The study, which was released in July 2004, concluded that it was feasible to send humans to Mars and that the costs would be roughly comparable to those of the Apollo program, adjusted for inflation.[55]

Once named Administrator, Griffin quickly diverged from the spiral development approach. On 13 June 2005, Griffin announced that Doug Cooke, a former DPT member and one of three Deputy Associate Administrators under Admiral Steidle, would replace the Admiral as head of the Exploration Systems Mission Directorate. Two weeks later, just days after Admiral Steidle's resignation became effective, Griffin made clear the reason for the move.[56] At a hearing of the House Committee on Science, Griffin was asked how he planned to ensure a consistent, enduring program of space exploration that would avoid the ups and downs, midcourse corrections, and cancellations of such previous programs as the ISS, the X-33, and the Orbital Space Plane. Griffin responded, in his typically blunt fashion, "You asked, what will we be doing different[ly]. First of all, I hope never again to let the words 'spiral development' cross my lips." After the laughter died down, Griffin explained that he felt spiral development was suitable for procuring certain large-scale military systems but was not relevant for NASA programs and that he preferred a "more direct approach." In practical terms, the more direct approach entailed doing away with the prolonged competitive process in the design phase, a central element of spiral development, and committing to a fixed design far earlier in order to accelerate development and reach program milestones sooner. Griffin also noted that he was looking forward to developing a direct plan to replace the Shuttle, create an architecture to send astronauts back to the Moon, and develop the Crew

55. "Extending Human Presence into the Solar System: An Independent Study for the Planetary Society on Strategy for the proposed U.S. Space Exploration Policy," July 2004, available at *https://web.archive.org/web/20080530123640/http://www.planetary.org/programs/projects/aim_for_mars/study-report.pdf* (accessed 13 June 2018). The Planetary Society also noted that at his first press conference as Administrator, Griffin fielded a question about sending astronauts to Mars by noting that he had co-led this study. See *https://web.archive.org/web/20080528224444/http://www.planetary.org/news/2005/0420_New_NASA_Administrator_Michael_Griffin.html* (accessed 13 June 2018).

56. Mark Carreau, "Exploration Chief Named as Key People Leave NASA," *Houston Chronicle* (14 June 2005).

Exploration Vehicle. He conceded that prioritizing the Moon and CEV would require postponing human exploration of Mars.[57]

Return to Flight and HST Servicing

While Administrator Griffin moved forward with a reshaped exploration agenda, Agency employees worked hard to rectify the specific causes of the Columbia accident. Since it was not possible to alter the fundamental design of the Space Transportation System, engineers and technicians did their best to improve their ability to identify and correct problems. Thus, foam could be applied to the external tank in ways that would make it less likely to shed on ascent, but at least some foam would continue to come off. Shuttle personnel improved visible light and infrared camera systems to track the vehicle after liftoff. On the way to the ISS, Shuttle crews would meticulously photograph the orbiter for ground personnel to determine if any thermal tiles had been damaged significantly. NASA managers also shifted launch schedules to have another "backup" Shuttle quickly available for a potential rescue mission if needed.

Although the technical causes of the Challenger and Columbia accidents were different, the process of ensuring flight safety for a return to flight after each of these fatal Space Shuttle accidents took two years. After careful planning and testing, the Shuttle returned to flight safely on 26 July 2005. Eileen Collins commanded the STS-114 mission aboard the Discovery orbiter. However, there were some concerns while Discovery was on orbit because detailed ascent imagery showed that a significant chunk of foam was shed from the external tank. Out of concern for safety, NASA delayed further Shuttle launches for 11 months, until Shuttle mission STS-121, on 4 July 2006.[58]

Prior to the successful resumption of Space Shuttle missions, one controversial aspect of returning the Shuttle to flight, and indeed of the VSE, was whether or not the Shuttle would be used for additional servicing missions for the Hubble

---

57. Michael Griffin testimony, "House Committee on Science Holds a Hearing on the Future of NASA," 28 June 2005, pp. 1, 3, 15–17, available at *https://www.nasa.gov/pdf/119619main_Griffin_Hil_testimony_062805.pdf* (accessed 7 May 2018). The quotations are from p. 16. In his opening statement, Congressman Bart Gordon commented, "At the same time, I must say that I am concerned about where NASA is headed and about the large number of unanswered questions that remain almost 18 months after the president announced his exploration initiative" (p. 3).

58. See the information at *http://www.nasa.gov/mission_pages/shuttle/shuttlemissions/archives/sts-114.html* (accessed 7 April 2011) and *http://www.nasa.gov/mission_pages/shuttle/shuttlemissions/list_2006.html* (accessed 9 May 2013). See, for example, John Logsdon, "NASA 2005—An Unfinished Play in Three Acts," *Space News* (19 December 2005) for a brief mention of the "major setback" caused by this foam shedding on STS-114.

Framed by Florida greenery, on 26 July 2005, Space Shuttle Discovery's STS-114 mission launches on the first Return to Flight mission since the loss of Space Shuttle Columbia on 1 February 2003. (NASA KSC-05PD-1737)

Space Telescope. During Sean O'Keefe's tenure as Administrator, the Columbia Accident Investigation Board questioned the safety of such missions since the Hubble's orbit was significantly different from that of the ISS and the Shuttle was incapable of making the orbital changes to get from one to the other. This meant that Shuttle astronauts could not take refuge on the ISS in the event that the Shuttle failed or experienced other serious problems during a Hubble servicing mission. Tasked with implementing the CAIB's decisions, the Return to Flight Task Group gained Administrator O'Keefe's consent to halt Hubble servicing missions. O'Keefe's primary concern was astronaut safety. Given his lack of technical experience, furthermore, he was not likely to reject the advice of experienced engineers or assume risks that he believed would ground the Shuttle fleet for good.

In the context of crafting the FY 2005 budget during the fall of 2003, while the VSE was simultaneously being formulated, Isakowitz, O'Keefe, and other policy-makers within NASA and the White House agreed not to include funding for a fifth Shuttle mission to service the Hubble. According to Isakowitz, this policy decision was driven by CAIB recommendations and more general budget discussions. In a later interview, he said "For those who…still argue that this was a budget decision, we cut the Hubble to pay for the vision, that is just simply not true. We would have found the money to do the Hubble." [59]

---

59. Steven J. Dick, "Appendix: The Decision To Cancel the Hubble Space Telescope Servicing Mission 4 (and Its Reversal)" in *Hubble's Legacy: Reflections by Those Who Dreamed It, Built It, and Observed the Universe with It*, ed. Roger D. Launius and David H. DeVorkin (Washington, DC: Smithsonian Institution Scholarly Press, 2014), pp. 151–189. The title of this appendix refers to Servicing Mission 4, which confusingly was actually the fifth servicing mission because the third mission was split into SM 3a and 3b. For the Isakowitz quote, see Steve Isakowitz, oral history by Steven Dick, 18 February 2004, pp. 10, 19, 25, 28, cited on p. 159 of this appendix essay.

On 18 May 2009, astronauts John Grunsfeld (left) and Andrew Feustel, both STS-125 mission specialists, participate in the mission's fifth and final session of extravehicular activity (EVA) to refurbish and upgrade the Hubble Space Telescope. (NASA s125e009603)

During the 19 December meeting with the President, O'Keefe explained that the HST servicing mission was off the table due to the CAIB recommendations, and President Bush concurred.[60] Immediately after President Bush's VSE announcement, however, news leaked that the VSE would not include a Hubble servicing mission. Two days later, O'Keefe told the Hubble team at NASA's Goddard Space Flight Center that he alone bore responsibility for the unpopular decision.[61] O'Keefe, Isakowitz, and Associate Administrator for (Human) Space Flight Bill Readdy all emphasized later that, as Isakowitz put it, "the Hubble decision really had…no specific link to the vision itself, but it was clear that if we were going to take a decision that said not to do it, it would cast a shadow on the vision."[62]

After considerable public, congressional, and internal NASA debate over the following two years and more, on 31 October 2006, Administrator Griffin

---

60. Sean O'Keefe, oral history by Steven Dick, 22 April 2004, pp. 3–6, cited in Dick, "Appendix: Decision To Cancel the HST SM4," pp. 161–162.

61. Dick, "Appendix: Decision to Cancel the HST SM4," p. 166.

62. Dick oral history with Isakowitz, 18 February 2004, p. 16, cited on p. 167 of Dick, "Appendix: Decision to Cancel the HST SM4."

overturned O'Keefe's decision: HST servicing missions would go forward. STS-125 launched on 11 May 2009 to service the Hubble for a fifth and final time.[63]

Shuttle program managers also decided to permit night launches after three successful daytime Shuttle launches in July 2005, July 2006, and September 2006. At the 28 November 2006 Flight Readiness Review for STS-116, the JSC Mission Operations Directorate reported that the Shuttle Program had concluded that launching during daylight hours "is highly desirable but *not a requirement* for future launches (post STS-115)."[64] Managers wanted to be able to see visible-light imagery of the orbiter ascending in case foam from the external tank struck the orbiter again. After several successful daytime launches, Shuttle managers believed that they were able to mitigate foam strikes reasonably well and that the risks of not being able to ascertain the cause of potential damage or "near misses" to the orbiter were outweighed by the need to fly a number of Shuttle missions to complete the ISS. Without night launches, launch personnel would be hard pressed to meet this schedule.[65] News accounts reported this decision to resume night launches as noncontroversial; one quoted Bill Gerstenmaier, NASA's Associate Administrator for Space Operations, as saying "There were really no dissenting opinions on the night launch."[66] The Shuttle Program was back on track for its successful completion, the first main goal of the VSE.

---

63. See Dick, "Appendix: Decision to Cancel the HST SM4," pp. 169–185 for a detailed accounting of this. See also, for example, *http://www.nasa.gov/mission_pages/shuttle/shuttlemissions/sts125/main/index.html* and *https://asd.gsfc.nasa.gov/archive/hubble/* (accessed 29 April 2011 and 13 June 2018, respectively).

64. John M. Curry, Anthony J. Ceccacci, and J. Stephen Stich, DA8, Flight Director Office, JSC Mission Operations Directorate, "STS-116/12A.1 Mission Operations," presentation for the STS-116 Flight Readiness Review, 28 November 2006, p. 56, DPT/VSE files, NASA HRC. Double emphasis is in the original.

65. Curry, Ceccacci, and Stich, "STS-116/12A.1 Mission Operations," pp. 55–60. See also the *Return to Flight Task Group Final Report: Assessing the Implementation of the Columbia Accident Investigation Board Return-to-Flight Recommendations* (Washington, DC: NASA, July 2005), available at *https://www.nasa.gov/pdf/125343main_RTFTF_final_081705.pdf* (accessed 13 June 2018), pp. 14, 54.

66. "NASA Gives Green Light to Shuttle Night Launch," *Associated Press* (30 November 2006). See "Night Shuttle Launch Will Not Prevent Debris Detection," *Space Daily* (6 November 2006), available at *http://www.spacedaily.com/reports/Night_Shuttle_Launch_Will_Not_Prevent_Debris_Detection_999.html* (accessed 10 August 2011), and Tariq Malik, "NASA: Space Shuttle Discovery Set for Dec. 7 Launch," Space.com, 29 November 2006, available at *http://www.space.com/3168-nasa-space-shuttle-discovery-set-dec-7-launch.html* (accessed 10 August 2011), for similar comments from Wayne Hale, the Shuttle Program Manager, and more on the launch constraints to finish building the ISS. Also see, 27 and 29 November and 9 November entries in Elaine Liston, *Chronology of KSC and KSC Related Events for 2006* (Washington, DC: NASA TM-2007-214727, February 2007) and *http://www.nasa.gov/mission_pages/shuttle/shuttlemissions/sts116/main/index.html* (accessed 14 April 2011).

8

CONCLUSION

EXPLORATION PLANNING that began at NASA before the arrival of Administrator Sean O'Keefe influenced the thinking of Agency officials who negotiated with the White House over the details of the VSE. The Decadal Planning Team and its successor planning group, the NASA Exploration Team, laid the groundwork for NASA's participation in the VSE development process. The experience greatly raised the level of knowledge of all aspects of exploration planning of a cadre of NASA scientists and engineers who later played critical roles in the VSE development process. Although he did not bargain directly with the White House over the contours of the VSE, Gary Martin, who served as the leader of NEXT and simultaneously as the NASA Space Architect, provided ideas, presentations, and various forms of support to those who did, including Steve Isakowitz, NASA's Comptroller and a trusted confidant of Sean O'Keefe.

Isakowitz was one of only three NASA officials (along with O'Keefe and his Chief of Staff, John Schumacher) deeply involved in negotiations with the White House over the VSE, and he was the only individual involved in all aspects of the process, including coordinating NASA proposals and responses to White House queries, negotiating with lower-level White House officials on the content of the VSE, building and defending budget options, and providing advice to Sean O'Keefe at critical junctures. In an interesting play on fate, Isakowitz also was one of two individuals responsible for initiating the Decadal Planning Team in 1999, when he served as the top civilian space budget examiner at the Office of Management and Budget. Although he did not attend the DPT/NEXT brainstorming and planning sessions, he was well aware of the teams' major ideas and products and worked closely with Gary Martin after he joined NASA in 2002. Through Gary Martin, Isakowitz harnessed the talents of NASA's leading scientists and engineers, many of whom participated in DPT or NEXT, to ensure that the concepts NASA put before the White House were sound and based on the most sophisticated Agency thinking about space exploration.

DPT and NEXT influenced the VSE in a direct sense in that Isakowitz and Martin drew upon DPT/NEXT concepts and charts in building presentations for senior NASA officials, including Sean O'Keefe, and officials in the White House. Many of the DPT/NEXT concepts, furthermore, were already embedded in Agency thinking at the highest levels. The Agency's 2003 strategic plan, for example, incorporated concepts that the DPT/NEXT teams advocated, such as using steppingstones, integrating human and robotic exploration, and basing decisions about exploration on broad scientific questions about the nature of the universe and human existence.[1] From the time he took office as NASA Administrator, Sean O'Keefe was aware of NEXT ideas and activities. He encouraged the activities to continue and later provided a means to integrate the DPT/NEXT ideas into NASA's formal plans by raising the visibility of Gary Martin and naming him the Agency's first Space Architect.

Despite the extensive planning work conducted at NASA between 1999 and 2003, the VSE differed in significant respects from both the proposals that emerged from DPT and NEXT before the Columbia accident and the proposals NASA made to White House officials throughout 2003. At the most basic level, the VSE reflected a compromise between NASA and several groups within the White House, including the Office of Management and Budget, the National Security Council, and the Office of Science and Technology Policy. NSC staff served as the arbiters of the process, with OMB, OSTP, and NASA as more or less equal participants. Officials from each organization argued vigorously over all aspects of the plan in the months before they presented a consensus proposal to the President in December 2003.

The consensus proposal did not favor one group or one set of ideas. All of the major participants left their mark. Yet, given NASA's position as the implementing agency, it played a central role in translating the general guidance from OMB, OSTP, and the NSC. OMB and the NSC, for example, led the charge to keep down total costs, but NASA determined how it would fund the early phases of the VSE (primarily by reallocating money from the Shuttle and ISS budgets as those programs wound down) and made difficult choices about which

---

1. National Aeronautics and Space Administration, *2003 Strategic Plan* (Washington, DC: NASA NP-2003-01-298-HQ, 2003), available at *https://www.nasa.gov/pdf/1968main_strategi.pdf* (accessed 26 May 2017).

programs to cut, shrink, or enlarge. Although OSTP pushed hard to make the Moon the focus of the VSE, NASA ensured that the President would frame extended operations on the Moon as a component of a more ambitious plan aimed at human exploration of the solar system. The President made his personal mark in asserting that Mars should be featured prominently in the VSE.

The Vision that the President announced, furthermore, included concepts that NASA promoted and introduced to the policy-making process, others that had multiple origins and supporters, and still others that NASA fought against vehemently. As mentioned earlier, the basic architecture of the plan, including using a steppingstones approach to gradually build capabilities to reach less accessible locations and aggressively integrating human and robotic capabilities, emerged from DPT/NEXT and found expression in NASA proposals to the White House and, ultimately, in the VSE.

As for ideas that had multiple supporters and origins, all participants in the process agreed that NASA needed to make tough decisions about program priorities to reduce costs and that budgetary considerations should determine the pace of the program. NASA and the White House suspected that the VSE would not gain support or remain sustainable in the long term if it garnered a large share of the national budget like the Apollo program or if total cost projections appeared too large, as occurred with SEI. For NASA, inspiration to follow a go-as-you-pay budgetary philosophy emerged from experience, but the history of the idea followed a path from the 1990 Augustine Commission Report, to the DPT Charter, to DPT/NEXT, and subsequently to all NASA proposals. Others involved in the process, including OMB and the NSC, drew the same conclusions from SEI and Apollo and advocated even more strongly than NASA for holding down initial cost projections.

As for elements of the VSE that NASA opposed, the most important and noteworthy was the method the White House compelled NASA to adopt to hold down costs. NASA supported a lean budget, particularly as an overarching approach to setting priorities and determining the pace of the program in the future. Yet NASA leaders argued vigorously against any choice that would limit the Agency's ability to fly humans in space continuously. With the support of Deputy Secretary of State Richard Armitage, Sean O'Keefe waged a campaign throughout the fall of 2003 to convince Deputy National Security Advisor Stephen Hadley and other high-level officials to support a near-term budget that would allow NASA to retire the Shuttle in 2010 and fly a replacement vehicle shortly thereafter. The refusal of Hadley and the White House to relent ultimately forced NASA to accept dependence on the Russian space program for access to space for approximately five years as a tradeoff for the VSE.[2]

Some of the ideas that DPT/NEXT promoted heavily prior to the Columbia accident, furthermore, never made it into the VSE or were watered down heavily. Science was not just one motivation among many for the exploration plans that DPT and NEXT conceived—science was the core "driver," meaning that DPT and NEXT expected scientific questions and goals to guide all decisions about exploration, including where to go and whether to use robots or humans. While internal NASA proposals continued to emphasize "science-driven" exploration in the months after the Columbia accident, science gradually declined from a critical framing concept for all NASA proposals to one of several spacefaring goals. Even before the White House initiated the final VSE policy planning process, NASA officials and planners began to emphasize other justifications for exploration, and they allowed questions about destinations, particularly the Moon, to gain primacy in their exploration plans. The consensus VSE presented to President Bush listed science as one of three major goals (along with extending the presence of humanity across the solar system and helping to promote U.S. technological innovation and leadership) and recommended holding down the budgets of NASA's Space Science and Earth Science programs to fund the VSE. With the NSC in control of the policy-making process, the concept of "science-driven" exploration vanished. The distant returns that science promised seemed to be insufficient to justify a long-term national commitment to space, particularly for a White House focused on immediate economic and national security concerns and committed to maintaining U.S. technological leadership.

As much as the ideas of DPT/NEXT informed the VSE, the fact that the President did not frame the VSE as a plan for "science-driven" exploration meant that the DPT/NEXT proposals remained an alternative to the Vision. The VSE, as announced and as implemented, focused on developing technology required to reach the Moon and to support an extended human presence on the Moon in preparation for human exploration of Mars and other destinations several decades in the future. Science was a secondary or tertiary consideration, and the sense of lunar exploration as a component of a larger plan to resolve fundamental scientific questions was absent. DPT/NEXT plans, importantly, included human exploration of the Moon, but as only one of several steppingstones. The

---

2. It is worth noting that the existing plan for ISS crew rescue already relied on Russia since OMB had canceled the X-38 Crew Return Vehicle as part of the "American Core Complete" restructuring after ISS costs had escalated. Thanks to Bill Barry for pointing this out. See also Krige, p. 182.

Moon's scientific potential, under DPT/NEXT plans, would determine the length of the stay and the extent of activities on the lunar surface.

In any event, policy planning followed a chronological continuum from DPT to NEXT to the VSE. At a very basic level, the goal was always to unshackle the next big step in human space exploration from the budgetary, technological, and policy limitations of the past. Planners wanted to find a way to send humans beyond low-Earth orbit, hopefully to Mars, by integrating human and robotic efforts in fundamentally new ways and overcoming perennial technological and budgetary roadblocks.

Whatever the ultimate differences between the DPT/NEXT plans and the VSE that emerged from the NSC-led interagency policy-making process, both NASA and White House officials felt strongly in the wake of the Columbia accident that the Agency needed to regain the confidence of Congress, the public, and international partners. Giving NASA a new mission vastly more ambitious than the path the space program had followed for the previous 30 years was not necessarily the obvious or appropriate response to an accident that had brought the capabilities of the Agency into question. Another presidential administration might have halted the Shuttle program entirely and directed the space agency not to fly humans in space until it developed a new vehicle. Other options that administration officials considered but then rejected included disbanding NASA and distributing its functions to other agencies or simply allowing the Agency to find its own way without presidential intervention.

In its own way, the fatal Columbia accident created an opportunity to take a fresh look at the nation's civilian space program. It was not clear what would result from the subsequent institutional soul-searching—whether NASA would be disbanded, emerge stronger, or something else. Policy-makers were careful not to be perceived as crassly opportunistic or insensitive, but they knew that this was an opportunity to forge a new direction for NASA and build upon work that DPT, NEXT, and other teams had done already. After the fact, O'Keefe stressed that the Columbia accident had provided a burning sense of urgency for all involved to arrive at a new strategic direction for NASA, where previously the situation had involved refinements of optional plans.[3] In short, without the loss of Columbia, the development of a new vision for human spaceflight likely would not have occurred in this time period.

The idea that the circumstances called for a major new initiative reached the status of conventional wisdom by the fall of 2003, with the Columbia Accident Investigation Board and several key members of Congress expressing support

---

3. O'Keefe oral history, passim and especially pp. 14–18.

for an overhaul of the human spaceflight program.[4] Yet the image President Bush projected of himself publicly—as a self-directed leader rarely influenced by Congress, experts, or conventional wisdom—would suggest that the President and his top advisors reached the conclusion on their own that the path from the Columbia accident should lead toward a stronger Agency with a more ambitious agenda. The fact that staff members in the Executive Office of the President had begun work on a new vision soon after the accident suggests strongly that the Bush administration would have considered a new mission for NASA with or without broader support from the public, Congress, or the space advocacy community.

The state of the human space program and the Bush administration's propensity to embark on bold policy initiatives in its first term seemed to favor a visible announcement of a broad presidential-level policy directive. While political calculations, including a desire to project an image of strength in the context of post–9/11 counterterrorism efforts and to avoid blame for the demise of the U.S. space program, may have contributed to the White House's decision, the administration might not have even considered a new policy if not for the existence of strong and longstanding sentiments that the Shuttle had fallen short of expectations and was nearing the end of its useful life.[5] As two of the key working-level White House staff members on space policy, Gil Klinger of the NSC and Brett Alexander of OSTP recognized immediately that the accident gave critics of human spaceflight an opportunity to argue for the termination of the Shuttle Program once and for all and to begin work on launch vehicles that could do more than ferry people and cargo to the ISS and back to Earth. But the White House could not simply terminate the Shuttle and start a new launch vehicle program without explaining how NASA would use the new vehicle. Ultimately, the former Texas governor and his administration looked at NASA with the same lenses as his father's administration, believing that the

---

4. Jeff Foust, "The Vision Thing," *Space Review* (10 November 2003).

5. Long before the second Shuttle accident, critics complained that the vehicles were outdated, were too expensive, and suffered from design flaws that could not be corrected. Critics also pointed to the risks of depending on a single type of launch vehicle for all payloads. See, for example, Hawthorne, Krauss & Associates, "Analysis of Potential Alternatives To Reduce NASA's Cost of Human Access to Space," 30 September 1998 (copy available in file 17588, NASA HRC); National Space Council, "Final Report to the President on the U.S. Space Program," January 1993; Thomas H. Johnson, "The Natural History of the Space Shuttle," *Technology and Society* 10 (1988): 417–244; Alex Roland, "Priorities in Space for the USA," *Space Policy* 3 (May 1987): 104–114; and even Robert C. Truax, "Shuttles—What Price Elegance?," *Astronautics and Aeronautics* (June 1970): 22–23, which was written before even the initiation of the Shuttle Program.

space program was not operating at its full potential if it lacked the ability to fly humans beyond LEO.

NASA was not fully prepared in a policy and planning sense for the Columbia accident. Although DPT/NEXT had been refining exploration concepts for four years, the assumptions for implementation did not include a chance event that would raise the possibility of openly initiating a program that included flying humans beyond LEO.[6] DPT began under Dan Goldin as a secretive effort aimed at developing plans for gradually increasing the space program's capabilities to the point where human missions beyond LEO would seem a natural, logical choice. With NEXT, the planning team became broader and more open, and its plans became more refined and concrete. The team, nonetheless, remained focused on identifying technological capabilities for enabling a broad range of human and robotic activities beyond LEO in the distant future. Sean O'Keefe encouraged the team to continue its work to feed the planning process, but he appears not to have had intentions of implementing anything so ambitious during his tenure. The accident, in a sense, halted the methodical, evolutionary approach to exploration that NASA had begun in 1999. Once the accident occurred, Sean O'Keefe and his inner circle of advisors reached the conclusion that the time had come to make the case for a more ambitious agenda.

In the end, did the Bush administration request enough funding for a major new direction in human spaceflight? Was an extra $1 billion spread over five years and the "reprogramming" of existing funding enough to start such an ambitious program?

Historical precedent suggests that the Bush VSE budget was severely inadequate. One benchmark for sending astronauts to the Moon and Mars might be the SEI program, which Aaron Cohen's 90-Day Study, in 1989, estimated would cost approximately $500 billion spread over 20–30 years to achieve.[7] Several years into the VSE program, the Senate Subcommittee on Science, Technology, and Space asked the Congressional Budget Office (CBO) to make

---

6. This is not to say that Dan Goldin and the DPT/NEXT teams were not aware of the possibility of an accident, only that Goldin had not initiated DPT/NEXT and the team had not conducted itself with any opportunistic intentions. As discussed in chapter 3, the expectation of both Goldin and the planning team was that implementation would occur slowly, out of the limelight, and without any dramatic shift for years and maybe even decades. As NASA Administrator, Goldin was deeply concerned about safety and was very concerned that another human spaceflight accident might occur on his watch as Administrator.

7. See the cost summary at *http://history.nasa.gov/90_day_cost_summary.pdf*. Adjusted for inflation, $500 billion in 1989 dollars would equate to approximately $766 billion in 2004 dollars or $1.01 trillion in 2018 dollars. These figures were calculated via the NASA New Start Inflation tool at *https://web.archive.org/web/20120902044105/http://cost.jsc.nasa.gov/inflation/nasa/inflateNASA.html* (accessed 7 November 2018).

its own assessment of the NASA and OMB cost estimates. Looking at historical cost growth on major programs such as Apollo and using different inflation figures from OMB's, the authors of the CBO report concluded that NASA would need an additional $32 billion through 2020, or approximately 12 percent above its projected budget and about 33 percent above the exploration portion.[8]

More importantly, although Congress supported the VSE policy goals, it was not willing to provide even the minimal amount the administration determined was needed to start the program off right. A historical analysis of presidential budget requests for NASA and the congressionally enacted levels shows that the former almost always have been higher than the latter. The trend continued in FY 2004–2006, when congressional appropriations were an average of $2.4 billion less than the President had requested. "The VSE became official policy, despite failing to gain any budgetary traction in Congress."[9]

When Griffin came aboard as NASA Administrator in spring 2005 (after spending time as the NASA Associate Administrator for Exploration during the SEI period), he anticipated that he would need to make hard choices about NASA's program priorities. According to John Logsdon, Griffin had "repeated over and over in his first six weeks in office [in spring 2005 that] there simply is not enough money in the NASA budget to do all meritorious things, and thus some worthwhile efforts will have to be deferred, given lower priority or not done at all." Furthermore, Griffin fully accepted the CAIB's point that many of NASA's problems since the end of the Apollo era had been the result of "straining to do too much with too little."[10]

In subsequent years, the Bush administration also seemed to lose enthusiasm for NASA and the VSE. The administration's FY 2006 NASA budget request included an increase that was only about half of what had been proposed in January 2004. OMB also rejected NASA's request for a 9 percent increase for FY 2007. With costs for the remaining Shuttle missions higher than initially expected, Griffin felt compelled to restrict funding for NASA's science priorities to sustain its human exploration agenda.[11] By the 2008 presidential election, it was obvious to leaders in the space policy community that a "fundamental mismatch" between NASA's budget and the VSE policy was hampering

8.   Congressional Budget Office, *The Budgetary Implications of NASA's Current Plans for Space Exploration*, April 2009, pp. xiii–xvi, cited in Richard S. Conley and Wendy Whitman Cobb, "Presidential Vision or Congressional Derision? Explaining Budgeting Outcomes for NASA, 1958–2008," *Congress & the Presidency* 39, no. 1 (2012): 57–58.

9.   Conley and Cobb, "Presidential Vision or Congressional Derision?" pp. 58–59, 66.

10. John Logsdon, "A Harder Road Ahead for Mike Griffin," *Space News* (6 June 2005).

11. John Logsdon, "NASA 2005—An Unfinished Play in Three Acts," *Space News* (19 December 2005).

implementation of the policy. As one respected analyst put it, the five-year budget that accompanied the VSE announcement was "woefully inadequate" for completing Shuttle operations, finishing the ISS, and sending astronauts back to the Moon by 2020.[12]

Free of the flaws that undermined George H. W. Bush's Space Exploration Initiative and advanced by an administration at the height of its power and influence, the VSE succeeded in helping the Agency move past the Columbia accident to place the human spaceflight program on a more stable foundation. Whatever difficulties the VSE posed for the Agency, it did free the human space program from what many considered the shackles of the Shuttle and ISS. It also renewed the idea of human spaceflight as the *raison d'être* of the Agency at a time when the future of NASA and its human spaceflight program was in question.

---

12. John Logsdon, "Time To Replace the Nixon Space Doctrine," *Space News* (29 September 2008).

Appendix A
A BRIEF NOTE ON SOURCES AND ACKNOWLEDGMENTS

AMONG THE ADVANTAGES of researching recent historical events are the accessibility of the people involved in the story and the availability of the written record they created. We were fortunate to have obtained copies of a wealth of relevant primary documents created as the events in this study unfolded, from 1999 to 2004, and to have conducted over one dozen oral histories shortly thereafter. All interview subjects provided useful insights, in part because the events in question were still fresh enough for them to remember clearly. As with all research materials for NASA History publications, all of the hard-copy and electronic source documents, including oral histories, have been deposited in the NASA Historical Reference Collection through the good graces of our esteemed colleagues, former NASA Chief Archivist Jane Odom and her successor, Robyn Rodgers.

This project began as a sponsored history of NASA's Decadal Planning Team and then became a two-part story that incorporated the policy development of the Vision for Space Exploration. All participants with whom we spoke, both formally in oral history interviews and informally in a variety of contexts and settings, were extraordinarily generous with their thoughts and time and greatly improved our understanding of these critical years in space history.

One area in which we were not as successful as we would have liked was in accessing White House documentation for the approximately one year between the Columbia accident and the public announcement of the VSE. We tried various channels, but we were denied access to official presidential records on the VSE policy development during the period in which we conducted research for this study. In addition, we were dismayed to learn that a top administration official had expressly decreed that handwritten notes from a key meeting should *not* be typed up and circulated within the White House; not only did this have a detrimental effect on the historical record, but it likely hindered successful policy implementation by allowing confusion about the intentions of the President to persist among his key advisors and those charged with implementing the VSE. We believe that the great breadth and depth of materials we received in other forms from key participants in DPT and the VSE policy development offset this gap in our knowledge and allowed us to write an insightful and valuable history.

It is fitting that this two-part story about policy formulation was a team effort. A number of former NASA History interns deserve special acknowledgment. A very special thanks to Mike Makara, who drafted an initial version of chapter 7 before going on to earn a Ph.D. in political science and becoming a faculty member at the University of Central Missouri. Liz Suckow prepared a detailed finding aid, and Amelia Lancaster consolidated it; Nicole Bucchino waded through speeches to find key quotations; Anna Stolitzka checked and updated many Web citations; and Will McCormick and Andres Almeida graciously and carefully reviewed the layout. Gabriel Okolski was most helpful in compiling and writing the first draft of the biographical appendix.

In the Office of International and Interagency Relations, Kent Bress was very gracious and supportive of this project, as was NASA's European representative, Tim Tawney; Kent also encouraged interns Jay Alver and John Rowley to help us, which they definitely did by finding multiple useful documents. Warm thanks to Dave Richards and Andrew Park for encouragement along the way and for quickly clearing an old document from export control.

Archivists Jane Odom, Colin Fries, John Hargenrader, and Liz Suckow provided research assistance and helpful advice over many years. Nadine Andreassen exceeded our expectations in providing administrative support and friendship along the way. Christian Gelzer's detailed and thoughtful review overturned many of our ideas of what this manuscript is about and how it might be received. Toward the end, Jonathan Krezel and Zach Pirtle at NASA Headquarters also provided key insights and assistance. Thanks also to the anonymous peer reviewers, who helped us strengthen the manuscript in various ways, and to our colleagues at the National Air and Space Museum and the Office of the Secretary of Defense (OSD) Historical Office, especially Thomas Lassman, Elliott V. Converse III, Roger Launius, and Michael Neufeld, who provided inspiration and insights that have shaped our thinking about history and space policy over the years.

Current NASA Chief Historian Bill Barry carefully reviewed the manuscript and supported us with much patience, as did former Chief Historian Steve Dick, who initiated this project at the suggestion of Harley Thronson of

Goddard Space Flight Center, who in turn showed great persistence in making sure this story got told. This history would not have been written without Harley's encouragement and support. Giulio Varsi deserves credit for his insights, patient encouragement, and tactical brilliance in helping us get past a number of roadblocks.

Thanks to our ace production team in the Communications Support Services Center (CSSC), who made this possible. Lisa Jirousek and Shawna Byrd smoothly copyedited the manuscript; Michele Ostovar did a wonderful job laying out the design; Tun Hla expertly handled the printing; and Maxine Aldred and Adriana Guevara oversaw the entire production team. Everybody there was a pleasure to work with and a true professional. Additional thanks to Kristin Harley, who did a skillful job of developing the index. Special thanks to Harley Thronson, Therese Griebel, and Jay Falker in the Space Technology Mission Directorate and to Randy Harris and Nadine Andreassen for their key roles in facilitating the printing.

Finally, a heartfelt thanks to Lynne, Josh, Naomi, Brej, Avery, Ayla, and all of our friends and family for their love and support.

Appendix B
BIOGRAPHICAL APPENDIX

BRETTON S. F. ALEXANDER spent five years as the senior policy analyst for space issues in the White House Office of Science and Technology Policy in the late Clinton and early George W. Bush administrations. Alexander holds bachelor's and master's degrees in aerospace engineering from the University of Virginia. He started his professional career in the Federal Aviation Administration's Office of Commercial Space Transportation. In the mid-1990s, Alexander lived in Moscow for more than a year and worked to facilitate space cooperation between Russia and the United States. He joined Transformational Space Corporation (t/Space) in January 2005 and became the company's vice president for government relations. He became the director for strategy and business development for Blue Origin in November 2011.

DOUGLAS COOKE worked on human exploration of space for most of his career and was a key DPT and NASA Exploration Team (NEXT) participant. He graduated from Texas A&M University in 1973 with a bachelor's degree in aerospace engineering. Having worked on the Space Station Freedom program and in the Space Shuttle Program Office, Cooke was an important part of planning for the 1989 Space Exploration Initiative. Cooke later served as NASA technical advisor to the Columbia Accident Investigation Board (CAIB). He joined the Exploration Systems Mission Directorate in January 2004 and then became its Deputy Associate Administrator in September 2005. He retired from NASA in August 2011 as the Associate Administrator for ESMD.

ROGER K. CROUCH provided an astronaut's perspective to the early DPT. After working as a visiting scientist at the Massachusetts Institute of Technology (MIT) from 1979 to 1980, he was Chief Scientist of the NASA Microgravity Space and Applications Division from 1985 to 1996. Crouch served as program scientist on five Spacelab flights. He logged more than 471 hours in space as a payload specialist on STS-83 (April 1997) and STS-94 (July 1997). Crouch also served at Headquarters as Senior Scientist for the Office of Life and Microgravity Sciences (1998–2000), Senior Scientist for the International Space Station Program (2000–2005), and University Affairs Officer for Space Exploration (2005–2006).

PETER CURRERI was one of two Marshall Space Flight Center representatives on DPT. He began his NASA career in 1981 as an expert in materials processing in space. He served as mission scientist for three Spacelab missions from 1993 to 1997, along with missions two, three, and four of the United States Microgravity Payload. He led the Biological and Physical Space Research Laboratory at Marshall from 2001 to 2004. He was the Lead Scientist for Exploration in the MSFC Exploration Science and Technology Division from 2005 to 2006. Curreri works in the In Situ Resource Utilization office at MSFC.

MARY S. DIJOSEPH was the executive director of DPT from February through August 2001 and later worked with NEXT as chair of the Human and Robotic Working Group through 2002. DiJoseph received her bachelor of science in mechanical engineering from MIT in 1985. She began her career at General Electric's Astro Space Division as an attitude control engineer. DiJoseph started at NASA's Goddard Space Flight Center in 1990 and became Deputy Program Manager for the Living With a Star Program in 2003. She oversaw the development of the Solar Dynamics Observatory (SDO) and the Space Environmental Testbeds (SET) project. She was selected as the Director of the Flight Projects Directorate at NASA's Langley Research Center in June 2014.

GUY FOGLEMAN was asked to be a part of DPT in the spring of 2000 to address human factors issues. He earned a bachelor of science in physics from Louisiana State University, a master of arts in mathematics and a master of science in physics from Indiana University, and a doctor of philosophy in physics from Indiana University. He has held positions as an associate professor of physics at San Francisco State University, as a visiting physicist in the Theory Group at the Stanford Linear Accelerator Center, and as a project scientist at NASA's Ames Research Center. From 2000 to 2004, he was the Director of the Bioastronautics Research Division in the Office of Biological and Physical

Research at NASA Headquarters. He then served for two years as the Associate Director for Human Health and Performance in the Exploration Systems Mission Directorate. Fogleman left NASA in 2006 and became the Executive Director of the Federation of American Societies for Experimental Biology. He retired in December 2016.

JAMES B. GARVIN was the first chair of DPT, from 1999 until 2001. Garvin received an undergraduate degree from Brown University in 1978 and a master of science from Stanford University in 1979. For doctoral work, he returned to Brown, where he received his doctor of philosophy in planetary geological sciences in 1984. In late 1984, Garvin began his NASA career at Goddard Space Flight Center, working to develop new remote sensing instruments such as the Shuttle Laser Altimeter. He also was a co-developer of the Earth System Science Pathfinder program. From 2000 to 2004, Garvin was NASA's first Chief Scientist for Mars Exploration. From 2004 to 2005, he served as Chief Scientist at NASA Headquarters. He became Chief Scientist of Goddard in 2005.

DANIEL S. GOLDIN commissioned DPT during his time as NASA's ninth and longest-serving Administrator, from April 1992 to November 2001. He graduated from the City College of New York with a mechanical engineering degree. He began his career in at NASA in 1962 as a research scientist at the Lewis (now Glenn) Research Center in Cleveland, Ohio. In the late 1960s, he went to work for the technology company TRW, where he rose to vice president and general manager of the Space and Technology Group in 1987. In 1992, President George H. W. Bush chose Goldin to occupy NASA's top spot, a position he held through the entire Clinton administration and for an initial part of President George W. Bush's administration. Goldin adopted a "faster, better, cheaper" philosophy, particularly for robotic space missions. He also cut the Agency's civil service workforce by about one-third without forcing layoffs. He was instrumental in helping to redesign the International Space Station Program and turning it from a plan into a reality. After leaving NASA, he founded KnuEdge (formerly known as the Intellisis Corporation) in 2005.

LISA GUERRA was one of the original members of DPT and served as its manager. Guerra earned a bachelor of science in aerospace engineering and a bachelor of arts in English from the University of Notre Dame and a master's degree in aerospace engineering from the University of Texas at Austin. Guerra began her professional career working at Eagle Engineering Corporation, where she focused on design studies of spacecraft for human missions to the Moon and Mars. She also worked at Science Applications International Corporation to support NASA's Johnson Space Center. At NASA Headquarters, she worked in the Exploration Systems Mission Directorate, the Biological and Physical Research Enterprise, and the Space Science Enterprise as special assistant to the Associate Administrator. In 2005, she was named the acting director of the Exploration Systems Mission Directorate's Integration Office. She is a Senior Advisor, Technical, to the NASA Administrator.

STEPHEN J. HADLEY served as the Deputy National Security Advisor in President George W. Bush's first term and oversaw the development of the VSE. Hadley is a graduate of Cornell University and Yale University Law School. Hadley's career has focused on defense and national security. He served under President George H. W. Bush as the Assistant Secretary of Defense for International Security Policy from 1989 to 1993. He also served as Counsel to the Tower Commission, a member of the National Security Council Staff, and an analyst for the Comptroller of the Department of Defense. From 2005 to 2009, he served as the National Security Advisor. After leaving government service, he cofounded the strategic consulting firm RiceHadleyGates, and he has served on the board of numerous public and private organizations, including the U.S. Institute of Peace.

G. SCOTT HUBBARD was an original member of DPT. He began his career conducting research in radiation-detection materials and devices. He was a staff scientist at the Lawrence Berkeley National Laboratory, a founder of the San Francisco Bay Area Canberra Semiconductor, and a senior research physicist at SRI International. In 1987, Hubbard came to NASA's Ames Research Center, where he worked his way up to a number of upper-level positions. He was the project manager for Ames's responsibilities for the Pathfinder mission that landed on Mars in 1997. In March 2000, Hubbard moved to NASA Headquarters, where he became the first Mars program director. He returned to Ames as the Center Director from 2002 to 2006 and also served as a member of the Columbia Accident Investigation Board. He is a professor of aeronautics and astronautics at Stanford University.

STEVE ISAKOWITZ provided "seed funding" for decadal planning from his post as OMB examiner and then also played a key role in developing the VSE from his job as the NASA Comptroller. He received his bachelor's and master's degrees in aeronautics and astronautics from MIT. He started his career at the management consulting firm Booz | Allen | Hamilton, where he analyzed opportunities for possible commercial space projects. Isakowitz worked at Lockheed Martin on several launch vehicle programs and later moved to the Office of

Management and Budget, where he became the Branch Chief of Science and Space Programs. In 2003, Isakowitz moved to his first position at NASA, serving as the Agency's Comptroller. In January 2005, he was appointed the Deputy Associate Administrator for the Exploration Systems Mission Directorate. He later served as the president of Virgin Galactic. In 2016, he became the president and chief executive officer of the Aerospace Corporation.

CHARLES "LES" JOHNSON served on DPT as an expert in space transportation concepts. He earned a bachelor's degree in chemistry and physics in 1984 from Transylvania University in Lexington, Kentucky, and a master's in physics in 1986 from Vanderbilt University in Nashville, Tennessee. He served as manager of the Science Programs and Projects Office at MSFC and then, from 2003 to 2005, as the manager of In-Space Transportation Technology at the Center's Advanced Space Transportation Program. He became Deputy Manager for the Advanced Concepts Office at NASA MSFC in 2008.

MARY E. KICZA served as the Associate Administrator for Biological and Physical Research and was part of a small "Rump Group" that helped develop the VSE. She received a bachelor's degree in electrical and electronics engineering from California State University, Sacramento, and a master's in business administration from the Florida Institute of Technology. Kicza began her career at NASA in 1982 at Kennedy Space Center as lead engineer for the Centaur Engineering Support Group. She held various positions at NASA Headquarters, including in the space science area, between 1987 and 1995. In 1995, she became the Associate Center Director for Space Science Programs at Goddard. She also served at Headquarters as the Associate Deputy Administrator for Systems Integration from August 2004 to May 2005. Her last NASA position was in 2005 as the acting Director for Business Management in the Science and Exploration Directorate at Goddard. In July 2014, she retired from her position as the Deputy Assistant Administrator for Satellite and Information Services at the National Oceanic and Atmospheric Administration.

GIL KLINGER was the National Security Council's Director of Space Policy from 2002 to 2005 and managed an interagency review of the nation's space policy, which became the VSE. Klinger received his undergraduate degree in European history and political science from the State University of New York at Albany in 1981 and received his master's in public policy from the John F. Kennedy School of Government at Harvard University. He served as Director for Space and Advanced Technology Strategy in the Office of the Under Secretary of Defense for Policy from 1991 to 1998. From 1998 until 2002, Klinger was the director of policy at the National Reconnaissance Office. He served from July 2005 to July 2009 as an assistant deputy director in the Acquisition and Collection Directorates of the Office of the Director of National Intelligence. He is vice president for space and intelligence at Raytheon Corporation.

GARY L. MARTIN was named NASA's Space Architect in October 2002 and also chaired NEXT. Martin earned a bachelor's degree in anthropology from Colorado State University, a bachelor's degree in applied math and physics from Virginia Commonwealth University, and a master's in mechanical engineering with an aerospace engineering concentration from the George Washington University. He served as the Chief of the NASA Technology Planning and Integration Office at Goddard Space Flight Center. He first worked at NASA Headquarters in 1990 as a program manager and Branch Chief in what was then known as Microgravity Sciences and Applications. In 2000, Martin became head of the Advanced Systems Office for the Office of Space Flight at NASA Headquarters. He is the Director of New Ventures and Communication at Ames Research Center.

SEAN O'KEEFE served as the 10th NASA Administrator, from 2001 to 2005. O'Keefe's tenure as Administrator saw the Columbia accident and the planning and announcement of the VSE. O'Keefe received his bachelor of arts in 1977 from Loyola University in New Orleans and his master of public administration in 1978 from the Maxwell School of Citizenship and Public Affairs at Syracuse University. From 1989 to 1992, he was the Comptroller and Chief Financial Officer of the Department of Defense. He served as Secretary of the Navy from 1992 until 1993. Prior to becoming NASA Administrator, O'Keefe was the Deputy Director of the Office of Management and Budget. After his tenure at NASA, he served as the chancellor of Louisiana State University. In 2014, O'Keefe stepped down as the chairman and chief executive officer of the Airbus Group's North American unit (formerly known as EADS North America) and became a distinguished senior advisor at the Center for Strategic and International Studies and a professor at Syracuse University.

DONALD R. PETTIT was the astronaut representative to the early DPT. He received a doctorate in chemical engineering from the University of Arizona in 1983. From 1984 to 1996, he worked as a staff scientist at Los Alamos National Laboratory, where he studied reduced-gravity fluid flow and materials processing experiments on board the NASA KC-135 aircraft. In 1990, after President George H. W. Bush unveiled his Space Exploration Initiative, Pettit was a member of the Synthesis Group tasked with investigating the technologies needed to return to the Moon and send humans to Mars. He served on the 1993 Space Station Freedom Redesign Team, which helped put the ISS in its form today. Selected as an astronaut in 1996, he first flew in space from November 2002 to May 2003, as science officer for ISS Expedition 6. His most recent trip to space was on Expedition 30/31 (December 2011 to July 2012).

DAVID RADZANOWSKI was a supervisory budget examiner at the Office of Management and Budget who covered NASA and space issues and helped develop the VSE. He began his career in space policy working at the Congressional Research Service. In 2006, he moved from OMB to NASA, taking a budget analysis job in the Space Operations Mission Directorate. From 2010 to 2014, he served as Chief of Staff to NASA Administrator Charles Bolden. He was NASA's Chief Financial Officer from 2014 to January 2017.

JOSEPH H. ROTHENBERG was NASA's Associate Administrator for Spaceflight from 1998 until 2001 and oversaw the work of DPT as one of two steering committee members. Rothenberg earned a bachelor of science in engineering science and a master of science in engineering management from C.W. Post College of Long Island University. He began his career in 1964 with Grumman Aerospace, where he worked on the Orbiting Astronomical Observatory and Solar Maximum Mission programs. Before joining Goddard Space Flight Center as Operations Manager for the Hubble Space Telescope, Rothenberg worked for Computer Technology Associates from 1981 to 1983. He was the Associate Director for Flight Projects for the Hubble Space Telescope program from 1990 to 1994, a period that saw the launch of the telescope and a servicing mission to fix the spacecraft's blurred optics. From 1995 to 1998, he served as Director of Goddard Space Flight Center. He retired from NASA in 2001.

MARK P. SAUNDERS was an original member of DPT. Saunders graduated from the Georgia Institute of Technology in 1970 with a degree in industrial engineering. He joined the U.S. Navy out of college and later worked as a civilian on the Navy's Poseidon/Trident missile submarine program. In 1989, Saunders began working at NASA on the Space Station Freedom program. Saunders later became the director of Langley Research Center's exploration and space access program, overseeing the development of new technologies for spacecraft and launch vehicles, among other projects. Saunders also worked at NASA Headquarters in the Office of Space Science, where he was put in charge of the office's Discovery Program, a low-cost solar system exploration initiative. He retired from NASA in 2008, after last serving as the Director of the Independent Program Assessment Office.

JOHN D. SCHUMACHER was the NASA Administrator's Chief of Staff from 2003 to 2005 and was involved in planning the VSE. A 1976 graduate of the United States Naval Academy with a bachelor's degree in oceanography and general engineering, Schumacher also obtained a master's degree in government from Georgetown University and a juris doctor degree from the Columbia University School of Law. He served at sea aboard the USS Guadalcanal from 1978 to 1982. He came to NASA in 1989 and served as the technical assistant to the Administrator and later as advisor to the Administrator. In 1991, Schumacher moved to the Office of External Relations, where he served briefly as the deputy and then the head of that office until 2003. In 2006, he became the vice president for Washington, DC, operations at Aerojet Corporation.

HARLEY A. THRONSON, JR., was a key member of DPT. Thronson received his bachelor's degree in astronomy from the University of California, Berkeley, in 1971 and his doctor of philosophy in astronomy and astrophysics from the University of Chicago in 1978. In 1996, after almost 20 years of teaching at the University of Wyoming in Laramie, he began working in NASA's Space Science Enterprise. He became a senior scientist in NASA Headquarters' Office of Space Science in 1999. In 2001, he began serving as the Office of Space Science's technology director. From 2004 to 2005, he served as the Assistant Associate Administrator of Technology for the Science Mission Directorate at NASA Headquarters. He became the Senior Scientist for Advanced Concepts in the Science and Exploration Directorate at NASA GSFC in July 2011.

EDWARD J. WEILER served as Associate Administrator for NASA Headquarters' Space Science Enterprise from 1998 to 2004 and oversaw the work of DPT as one of two steering committee members. Weiler earned a doctorate in astronomy in 1976 from Northwestern University. In 1978, he joined NASA as a staff scientist at Headquarters and was promoted to chief of the Ultraviolet/Visible and Gravitational Astrophysics Division the following year. From 1979 to 1998, Weiler served as Chief Scientist for the Hubble Space Telescope program. He served as Director of Goddard Space Flight Center from August 2004 to May 2008. From 2008 until his retirement from NASA in 2011, he led NASA's Science Mission Directorate.

JOSEPH R. WOOD served as Special Advisor for National Security Affairs to Vice President Dick Cheney from 2001 to 2003; in this position, he was responsible for space issues and helped develop the VSE. Wood is a graduate of the United States Air Force Academy and holds a master's degree in public administration from Harvard University's Kennedy School of Government. In March 2004, he was appointed Deputy Assistant Administrator for NASA's Office of External Relations, and he helped coordinate U.S. government efforts to engage international partners in cooperative space activities. A retired Air Force colonel, he is currently a professor at the Institute of World Politics.

Appendix C
CHRONOLOGY

FALL 1998–SPRING 1999: OMB passback for FY 2000 included funding for decadal planning. OMB examiners Doug Comstock and Steve Isakowitz put a line item for decadal planning in NASA's FY 2000 budget.

4 APRIL 1999: NASA Administrator Dan Goldin held a meeting at his house to initiate DPT.

1 JUNE 1999: NASA Associate Administrators Ed Weiler and Joe Rothenberg formally set up DPT with Jim Garvin and Lisa Guerra as leaders.

JUNE–DECEMBER 1999: DPT Phase 1 occurred.

24–25 JUNE 1999: The first DPT meeting was held in Washington, DC.

JANUARY–OCTOBER 2000: DPT Phase 2 (design studies to validate capabilities) occurred.

12–13 OCTOBER 2000: Wye River, Maryland, briefing to Goldin took place (ISS cost overruns discussed as big problem then).

31 OCTOBER 2000: ISS Expedition 1 launched.

JANUARY 2001–FEBRUARY 2002: DPT Phase 3 (NEXT) occurred—Gary Martin took over leadership.

11 SEPTEMBER 2001: Terrorists attacked World Trade Center and Pentagon.

7 OCTOBER 2001: Operation Enduring Freedom began.

29 OCTOBER 2001: Dr. John Marburger became presidential Science Advisor.

17 NOVEMBER 2001: Goldin left NASA.

21 DECEMBER 2001: Sean O'Keefe became NASA Administrator.

APRIL 2002: Isakowitz became NASA Comptroller. Comstock joined Isakowitz at NASA, and David Radzanowski took over as OMB Science and Space Programs Branch Chief.

AUGUST 2002: Codes S, U, and M signed Memo of Agreement to formally create NEXT.

11 OCTOBER 2002: Martin was named NASA Space Architect, reporting to Deputy Administrator Fred Gregory.

1 FEBRUARY 2003: Columbia Space Shuttle accident occurred.

MARCH 2003: President Bush launched Operation Iraqi Freedom.

EARLY SUMMER 2003: White House formed "Rump Group" with John Schumacher, Steve Isakowitz, and Mary Kicza from NASA.

LATE SUMMER–FALL 2003: Staff-level meetings were held in parallel with Deputies Committee meetings.

FALL 2003: Project Prometheus was included in the President's 2004 budget.

17 DECEMBER 2003: Centennial of Flight anniversary took place.

19 DECEMBER 2003: Briefing of the President was held to finalize the VSE.

14 JANUARY 2004: President George W. Bush gave the Vision speech at NASA Headquarters.

Appendix D
KEY DOCUMENTS

D1. DPT Charter

Charter

Decadel [*sic*] Planning Team (DPT)

Background
The Office of Management and Budget has set aside a small amount of yearly funds, starting in FY 2000, for NASA to: "Explore and refine concepts and technologies that are critical to developing a robust set of civil space initiatives at different funding levels for the next decade". This coincides with the NASA Administrator's direction to develop and [*sic*] overarching *Agency* plan for the first quarter of the New Millennium likewise dealing with scenarios and supporting technologies for space exploration. Decadal planning is defined as the first ten-year definitive planning to reach a twenty[-]five-year vision.

Composition
The DPT will be a small team (about 15) NASA civil servants and JPL employees. It will be selected by and report to a Steering Committee co-chaired by the Associate Administrators for Space Flight and Space Science.

Charge

The DPT will develop a vision of Space Exploration for the first quarter of the New Millennium (and perhaps, beyond). The vision should include the following characteristics:

1. top down
2. forward looking and NOT tied to past concepts
3. Science driven technology enabled program development approach with technology roadmaps that enable capabilities at an affordable cost
4. *aggressively* integrates robotic and human capabilities
5. opens the Human frontier beyond LEO by building infrastructure robotically at strategic outposts—L2, L1, planetary moons, planets, etc.
6. includes a wide-range of exploration tools (e.g., space planes, balloons, L1/2 human constructed and maintained observatories, etc.)
7. incremental (buy by the yard) as budget supports
8. propulsion system requir[e]ments driven by mission approaches
9. industry has commercialized LEO and GEO—NASA's job is to expand the frontier beyond earth orbit

The DPT will suggest a number of scenarios to achieve the above. These may be characterized by annual funding required or the schedule for specific major accomplishments within the vision. The team will take due cognizance of all of the past and ongoing activities initiated towards this and similar objectives and it will define major elements of the vision including safety drivers, resources, infrastructure, environment, health maintenance, mission models, and transportation requirements.

> **This study should be viewed as the first small step toward a program designed to enable the inevitable and systematic migration of humans and robots into space beyond earth orbit for the purposes of exploration, science and commerce.**

Duration

The DPT will complete Phase I of this effort. Phase I will provide a high level first cut [at] a range of scenarios across annual funding, accomplishments and breadth of activities from which potential follow-on study options can be selected. It is anticipated that Phase I will be concluded by October 1, 1999. The composition of the members for subsequent phases is TBD.

Authority

The DPT is to operate for the expected duration of its charter without special funding. It may request support from part of NASA without compensation to the organizations involved. *Support form [sic] outside NASA may be requested and accepted without NASA compensation.*

Operation

To facilitate and expedite the timely functioning of the DPT, a virtual environment using collaborative [*sic*] will be established by LaRC. This will provide the tools for Phase 1 and in subsequent Phases test the concepts of a fully implemented Intelligent Synthesis Environment (ISE). The DPT will actively interface with the Steering Committee as required on an informal basis and will hold scheduled discussions as determined by the progress of the work.

Deliverables

At the completion of Phase I, the DPT will deliver to the Steering Committee and to the Administrator the following:

    a. The vision for robotic and human exploration in the first quarter of the New Millennium.

    b. Various scenarios with "first order" required investment and schedule to realize the vision.

    c. A definition of critical technologies in each of the "resource areas" defined in the Terms of Reference for Decadel [*sic*] Study Team that needs to be funded now.

    d. A recommendation for follow-on phases.

Attachment

Terms of Reference for Decadel [*sic*] Study Team

A.A. for Space Flight

A.A. for Space Science

Date

Date

D2. DPT Team Members

James Garvin (GSFC), chair
Lisa Guerra (Headquarters)
Harley Thronson (Headquarters)
Roger Crouch (Headquarters)
Matt Golombek (JPL)
Barbara Wilson (JPL)
Mark Pine (JPL)
Dennis Bushnell (LaRC)
Mark Saunders (LaRC)*
Alan Wilhite (LaRC)
Dave Dawson (JSC)
Julie Kramer (JSC)*
Don Pettit (JSC)
Les Johnson (MSFC)
Peter Curreri (MSFC)
Lynn Harper (ARC)*
Peter Norvig (ARC)*
Scott Hubbard (ARC)
Jerome Bennet (GSFC)
Paul Westmeyer (GSFC)

* Indicates people who were not listed at the 24 June 1999 meeting

D3. Grand Challenges

Huntress's Grand Challenges
1. Read the history and destiny of the solar system.
2. Look for life elsewhere in the solar system.
3. Image and study extrasolar planets.
4. Send a spacecraft to a nearby star.
5. Conduct a progressive and systematic program of human exploration beyond Earth orbit.

Weiler's Grand Challenges
1. How did the universe begin?
2. How did we get here?
3. Where are we going?
4. Are we alone?

D4. Strategy Based on Long-Term Affordability

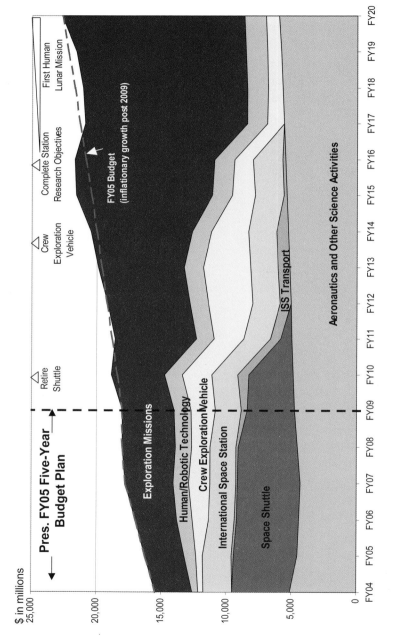

Source: This "sand chart" is available at *http://history.nasa.gov/sepbudgetchart.pdf*.

Appendix E
NASA ORGANIZATIONAL RESTRUCTURING

FOLLOWING THE UNVEILING of the VSE policy, NASA Headquarters adopted a new organizational structure that distinguished core mission activities from mission support activities and visibly raised the status of mission activities by consolidating them into just four major directorates: Exploration Systems, Space Operations, Science, and Aeronautics Research. An announcement went out on 24 June 2004 (*https://www.nasa.gov/home/hqnews/2004/jun/HQ_04205_Transformation.html*) to explain the reorganization in conjunction with the release of a "Clarity Team Report" of the same date (*https://www.hq.nasa.gov/office/hqlibrary/documents/o56097162.pdf*). Thanks to Annette Frederick and Nanette Jennings of NASA Headquarters for providing these background sources. See the two organization charts below for a "before" and "after" depiction of the organizational structure. The parenthetical letters in the "before" organizational chart are the "codes."

National Aeronautics and Space Administration
Office of the Administrator

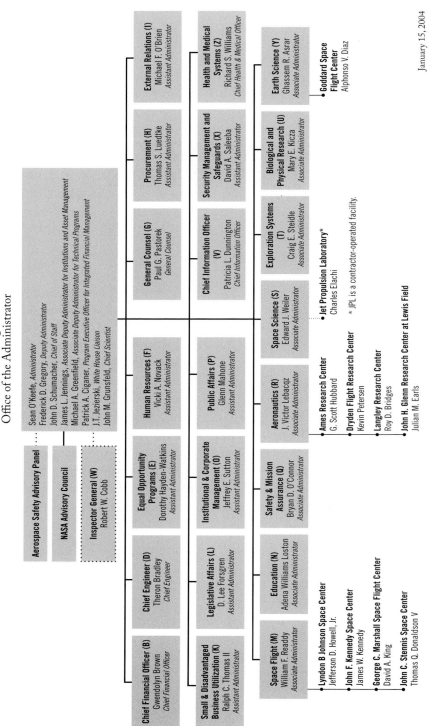

January 15, 2004

Aerospace Safety Advisory Panel

NASA Advisory Council

Inspector General (W)
Robert W. Cobb

Sean O'Keefe, *Administrator*
Frederick D. Gregory, *Deputy Administrator*
John D. Schumacher, *Chief of Staff*
James L. Jennings, *Associate Deputy Administrator for Institutions and Asset Management*
Michael A. Greenfield, *Associate Deputy Administrator for Technical Programs*
Patrick A. Ciganer, *Program Executive Officer for Integrated Financial Management*
J.T. Jezierski, *White House Liaison*
John M. Grunsfeld, *Chief Scientist*

Chief Financial Officer (B)
Gwendolyn Brown
Chief Financial Officer

Chief Engineer (D)
Theron Bradley
Chief Engineer

Equal Opportunity Programs (E)
Dorothy Hayden-Watkins
Assistant Administrator

Human Resources (F)
Vicki A. Novack
Assistant Administrator

General Counsel (G)
Paul G. Pastorek
General Counsel

Procurement (H)
Thomas S. Luedtke
Assistant Administrator

External Relations (I)
Michael F. O'Brien
Assistant Administrator

Small & Disadvantaged Business Utilization (K)
Ralph C. Thomas II
Assistant Administrator

Legislative Affairs (L)
D. Lee Forsgren
Assistant Administrator

Institutional & Corporate Management (O)
Jeffrey E. Sutton
Assistant Administrator

Public Affairs (P)
Glenn Mahone
Assistant Administrator

Chief Information Officer (V)
Patricia L. Dunnington
Chief Information Officer

Security Management and Safeguards (X)
David A. Saleeba
Assistant Administrator

Health and Medical Systems (Z)
Richard S. Williams
Chief Health & Medical Officer

Education (N)
Adena Williams Loston
Associate Administrator

Safety & Mission Assurance (Q)
Bryan D. O'Connor
Associate Administrator

Aeronautics (R)
J. Victor Lebacqz
Associate Administrator

Space Science (S)
Edward J. Weiler
Associate Administrator

Exploration Systems (T)
Craig E. Steidle
Associate Administrator

Biological and Physical Research (U)
Mary E. Kicza
Associate Administrator

Earth Science (Y)
Ghassem R. Asrar
Associate Administrator

Space Flight (M)
William F. Readdy
Associate Administrator

● **Lyndon B Johnson Space Center**
Jefferson D. Howell, Jr.

● **John F. Kennedy Space Center**
James W. Kennedy

● **George C. Marshall Space Flight Center**
David A. King

● **John C. Stennis Space Center**
Thomas Q. Donaldson V

● **Ames Research Center**
G. Scott Hubbard

● **Dryden Flight Research Center**
Kevin Petersen

● **Langley Research Center**
Roy D. Bridges

● **John H. Glenn Research Center at Lewis Field**
Julian M. Earls

● **Jet Propulsion Laboratory***
Charles Elachi

* JPL is a contractor-operated facility.

● **Goddard Space Flight Center**
Alphonso V. Diaz

Transformed Structure

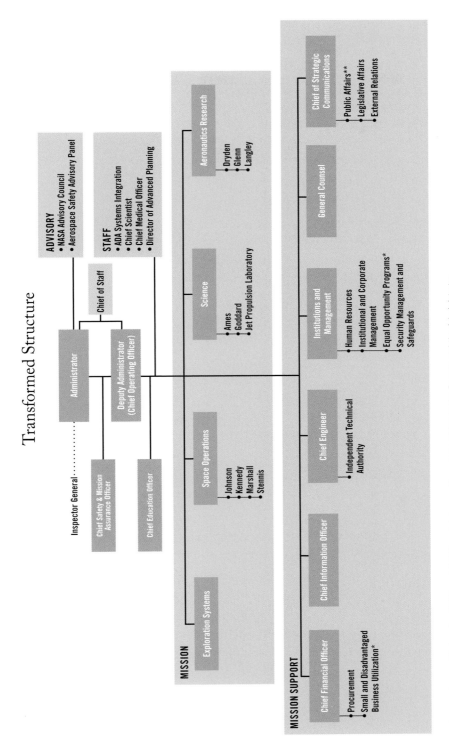

June 24, 2004

* In accordance with law, the OEOP and SDBU maintain reporting relationships to the Deputy and the Administrator.

** Including a new emphasis on internal communications

Appendix F
ACRONYMS

AIAA	American Institute of Aeronautics and Astronautics
ARC	Ames Research Center
CAIB	Columbia Accident Investigation Board
CBO	Congressional Budget Office
CEA	Council of Economic Advisors
CEV	Crew Exploration Vehicle
CIA	Central Intelligence Agency
CNSA	China National Space Administration
CSA	Canadian Space Agency
DARPA	Defense Advanced Research Projects Agency
DLR	*Deutsches Zentrum für Luft- und Raumfahrt e.V.* (German Aerospace Center)
DOD	Department of Defense
DPC	Domestic Policy Council
DPT	Decadal Planning Team
DSRI	Danish Space Research Institute
EADS	European Aeronautic Defence and Space Company
EELV	evolved expendable launch vehicle

EMPIRE	Early Manned Planetary-Interplanetary Roundtrip Expeditions
EOP	Executive Office of the President
ESA	European Space Agency
ESMD	Exploration Systems Mission Directorate
EVA	extravehicular activity
FAA	Federal Aviation Administration
FBC	faster, better, cheaper
FFRDC	Federally Funded Research and Development Center
FY	fiscal year
GAO	Government Accountability Office
GE	General Electric
GEO	geosynchronous orbit
GSFC	Goddard Space Flight Center
HEDS	Human Exploration and Development of Space
HEO	high-Earth orbit
HLR	Human Lunar Return
HRC	Historical Reference Collection
HRET	Human-Robotic Exploration Team

HST	Hubble Space Telescope	**NRO**	National Reconnaissance Office
ISS	International Space Station	**NSC**	National Security Council
ISTP	Integrated Space Transportation Plan	**NSF**	National Science Foundation
JAG	Joint Action Group	**NSPD**	National Security Presidential Directive
JAXA	Japan Aerospace Exploration Agency	**NSS**	National Space Society
JIMO	Jupiter Icy Moons Orbiter	**OER**	Office of External Relations
JPDO	Joint Planning and Development Office	**OMB**	Office of Management and Budget
JPL	Jet Propulsion Laboratory	**OSD**	Office of the Secretary of Defense
JSAC	Joint Strategic Assessment Committee	**OSP**	Orbital Space Plane
JSC	Johnson Space Center	**OSS**	Office of Space Science
KSC	Kennedy Space Center	**OSTP**	Office of Science and Technology Policy
LaRC	Langley Research Center	**PMRG**	Planetary Mission Requirements Group
LEO	low-Earth orbit	**R&D**	research and development
MIT	Massachusetts Institute of Technology	**RFP**	Request for Proposals
MOA	Memorandum of Agreement	**RTG**	Radioisotope Thermoelectric Generator
MRM	Mars Reference Mission	**RTQ**	Responses to Questions
MSFC	Marshall Space Flight Center	**SDO**	Solar Dynamics Observatory
MTPE	Mission to Planet Earth	**SEI**	Space Exploration Initiative
NAC	NASA Advisory Council	**SET**	Space Environmental Testbeds
NACA	National Advisory Committee for Aeronautics	**SLI**	Space Launch Initiative
NASA	National Aeronautics and Space Administration	**SOHO**	Solar and Heliospheric Observatory
NEAR	Near Earth Asteroid Rendezvous	**SOK**	Sean O'Keefe
NERVA	Nuclear Engine for Rocket Vehicle Applications	**STEM**	science, technology, engineering, and mathematics
NEXT	NASA Exploration Team; NASA Evolutionary Xenon Thruster	**STS**	Space Transportation System (Space Shuttle)
NGLT	Next Generation Launch Technology	**t/Space**	Transformational Space Corporation
NIST	National Institute of Standards and Technology	**VfR**	*Verein für Raumschiffahrt* (German Society for Space Travel)
NOAA	National Oceanic and Atmospheric Administration	**VSE**	Vision for Space Exploration
NRC	National Research Council	**WIRE**	Wide-Field Infrared Explorer
		WMAP	Wilkinson Microwave Anisotropy Probe

THE NASA HISTORY SERIES

Reference Works, NASA SP-4000

Grimwood, James M. *Project Mercury: A Chronology.* NASA SP-4001, 1963.

Grimwood, James M., and Barton C. Hacker, with Peter J. Vorzimmer. *Project Gemini Technology and Operations: A Chronology.* NASA SP-4002, 1969.

Link, Mae Mills. *Space Medicine in Project Mercury.* NASA SP-4003, 1965.

Astronautics and Aeronautics, 1963: Chronology of Science, Technology, and Policy. NASA SP-4004, 1964.

Astronautics and Aeronautics, 1964: Chronology of Science, Technology, and Policy. NASA SP-4005, 1965.

Astronautics and Aeronautics, 1965: Chronology of Science, Technology, and Policy. NASA SP-4006, 1966.

Astronautics and Aeronautics, 1966: Chronology of Science, Technology, and Policy. NASA SP-4007, 1967.

Astronautics and Aeronautics, 1967: Chronology of Science, Technology, and Policy. NASA SP-4008, 1968.

Ertel, Ivan D., and Mary Louise Morse. *The Apollo Spacecraft: A Chronology, Volume I, Through November 7, 1962.* NASA SP-4009, 1969.

Morse, Mary Louise, and Jean Kernahan Bays. *The Apollo Spacecraft: A Chronology, Volume II, November 8, 1962–September 30, 1964.* NASA SP-4009, 1973.

Brooks, Courtney G., and Ivan D. Ertel. *The Apollo Spacecraft: A Chronology, Volume III, October 1, 1964–January 20, 1966.* NASA SP-4009, 1973.

Ertel, Ivan D., and Roland W. Newkirk, with Courtney G. Brooks. *The Apollo Spacecraft: A Chronology, Volume IV, January 21, 1966–July 13, 1974.* NASA SP-4009, 1978.

Astronautics and Aeronautics, 1968: Chronology of Science, Technology, and Policy. NASA SP-4010, 1969.

Newkirk, Roland W., and Ivan D. Ertel, with Courtney G. Brooks. *Skylab: A Chronology.* NASA SP-4011, 1977.

Van Nimmen, Jane, and Leonard C. Bruno, with Robert L. Rosholt. *NASA Historical Data Book, Volume I: NASA Resources, 1958–1968.* NASA SP-4012, 1976; rep. ed. 1988.

Ezell, Linda Neuman. *NASA Historical Data Book, Volume II: Programs and Projects, 1958–1968.* NASA SP-4012, 1988.

Ezell, Linda Neuman. *NASA Historical Data Book, Volume III: Programs and Projects, 1969–1978.* NASA SP-4012, 1988.

Gawdiak, Ihor, with Helen Fedor. *NASA Historical Data Book, Volume IV: NASA Resources, 1969–1978.* NASA SP-4012, 1994.

Rumerman, Judy A. *NASA Historical Data Book, Volume V: NASA Launch Systems,*

Space Transportation, Human Space-flight, and Space Science, 1979–1988. NASA SP-4012, 1999.

Rumerman, Judy A. *NASA Historical Data Book, Volume VI: NASA Space Applications, Aeronautics and Space Research and Technology, Tracking and Data Acquisition/Support Operations, Commercial Programs, and Resources, 1979–1988.* NASA SP-4012, 1999.

Rumerman, Judy A. *NASA Historical Data Book, Volume VII: NASA Launch Systems, Space Transportation, Human Spaceflight, and Space Science, 1989–1998.* NASA SP-2009-4012, 2009.

Rumerman, Judy A. *NASA Historical Data Book, Volume VIII: NASA Earth Science and Space Applications, Aeronautics, Technology, and Exploration, Tracking and Data Acquisition/Space Operations, Facilities and Resources, 1989–1998.* NASA SP-2012-4012, 2012.

No SP-4013.

Astronautics and Aeronautics, 1969: Chronology of Science, Technology, and Policy. NASA SP-4014, 1970.

Astronautics and Aeronautics, 1970: Chronology of Science, Technology, and Policy. NASA SP-4015, 1972.

Astronautics and Aeronautics, 1971: Chronology of Science, Technology, and Policy. NASA SP-4016, 1972.

Astronautics and Aeronautics, 1972: Chronology of Science, Technology, and Policy. NASA SP-4017, 1974.

Astronautics and Aeronautics, 1973: Chronology of Science, Technology, and Policy. NASA SP-4018, 1975.

Astronautics and Aeronautics, 1974: Chronology of Science, Technology, and Policy. NASA SP-4019, 1977.

Astronautics and Aeronautics, 1975: Chronology of Science, Technology, and Policy. NASA SP-4020, 1979.

Astronautics and Aeronautics, 1976: Chronology of Science, Technology, and Policy. NASA SP-4021, 1984.

Astronautics and Aeronautics, 1977: Chronology of Science, Technology, and Policy. NASA SP-4022, 1986.

Astronautics and Aeronautics, 1978: Chronology of Science, Technology, and Policy. NASA SP-4023, 1986.

Astronautics and Aeronautics, 1979–1984: Chronology of Science, Technology, and Policy. NASA SP-4024, 1988.

Astronautics and Aeronautics, 1985: Chronology of Science, Technology, and Policy. NASA SP-4025, 1990.

Noordung, Hermann. *The Problem of Space Travel: The Rocket Motor.* Edited by Ernst Stuhlinger and J. D. Hunley, with Jennifer Garland. NASA SP-4026, 1995.

Gawdiak, Ihor Y., Ramon J. Miro, and Sam Stueland. *Astronautics and Aeronautics, 1986–1990: A Chronology.* NASA SP-4027, 1997.

Gawdiak, Ihor Y., and Charles Shetland. *Astronautics and Aeronautics, 1991–1995: A Chronology.* NASA SP-2000-4028, 2000.

Orloff, Richard W. *Apollo by the Numbers: A Statistical Reference.* NASA SP-2000-4029, 2000.

Lewis, Marieke, and Ryan Swanson. *Astronautics and Aeronautics: A Chronology, 1996–2000.* NASA SP-2009-4030, 2009.

Ivey, William Noel, and Marieke Lewis. *Astronautics and Aeronautics: A Chronology, 2001–2005.* NASA SP-2010-4031, 2010.

Buchalter, Alice R., and William Noel Ivey. *Astronautics and Aeronautics: A Chronology, 2006*. NASA SP-2011-4032, 2010.

Lewis, Marieke. *Astronautics and Aeronautics: A Chronology, 2007*. NASA SP-2011-4033, 2011.

Lewis, Marieke. *Astronautics and Aeronautics: A Chronology, 2008*. NASA SP-2012-4034, 2012.

Lewis, Marieke. *Astronautics and Aeronautics: A Chronology, 2009*. NASA SP-2012-4035, 2012.

Flattery, Meaghan. *Astronautics and Aeronautics: A Chronology, 2010*. NASA SP-2013-4037, 2014.

Siddiqi, Asif A. *Beyond Earth: A Chronicle of Deep Space Exploration, 1958–2016*. NASA SP-2018-4041, 2018.

Management Histories, NASA SP-4100

Rosholt, Robert L. *An Administrative History of NASA, 1958–1963*. NASA SP-4101, 1966.

Levine, Arnold S. *Managing NASA in the Apollo Era*. NASA SP-4102, 1982.

Roland, Alex. *Model Research: The National Advisory Committee for Aeronautics, 1915–1958*. NASA SP-4103, 1985.

Fries, Sylvia D. *NASA Engineers and the Age of Apollo*. NASA SP-4104, 1992.

Glennan, T. Keith. *The Birth of NASA: The Diary of T. Keith Glennan*. Edited by J. D. Hunley. NASA SP-4105, 1993.

Seamans, Robert C. *Aiming at Targets: The Autobiography of Robert C. Seamans*. NASA SP-4106, 1996.

Garber, Stephen J., ed. *Looking Backward, Looking Forward: Forty Years of Human Spaceflight Symposium*. NASA SP-2002-4107, 2002.

Mallick, Donald L., with Peter W. Merlin. *The Smell of Kerosene: A Test Pilot's Odyssey*. NASA SP-4108, 2003.

Iliff, Kenneth W., and Curtis L. Peebles. *From Runway to Orbit: Reflections of a NASA Engineer*. NASA SP-2004-4109, 2004.

Chertok, Boris. *Rockets and People, Volume I*. NASA SP-2005-4110, 2005.

Chertok, Boris. *Rockets and People: Creating a Rocket Industry, Volume II*. NASA SP-2006-4110, 2006.

Chertok, Boris. *Rockets and People: Hot Days of the Cold War, Volume III*. NASA SP-2009-4110, 2009.

Chertok, Boris. *Rockets and People: The Moon Race, Volume IV*. NASA SP-2011-4110, 2011.

Laufer, Alexander, Todd Post, and Edward Hoffman. *Shared Voyage: Learning and Unlearning from Remarkable Projects*. NASA SP-2005-4111, 2005.

Dawson, Virginia P., and Mark D. Bowles. *Realizing the Dream of Flight: Biographical Essays in Honor of the Centennial of Flight, 1903–2003*. NASA SP-2005-4112, 2005.

Mudgway, Douglas J. *William H. Pickering: America's Deep Space Pioneer*. NASA SP-2008-4113, 2008.

Wright, Rebecca, Sandra Johnson, and Steven J. Dick. *NASA at 50: Interviews with NASA's Senior Leadership*. NASA SP-2012-4114, 2012.

Project Histories, NASA SP-4200

Swenson, Loyd S., Jr., James M. Grimwood, and Charles C. Alexander. *This New Ocean: A History of Project Mercury*. NASA SP-4201, 1966; rep. ed. 1999.

Green, Constance McLaughlin, and Milton Lomask. *Vanguard: A History*.

NASA SP-4202, 1970; rep. ed. Smithsonian Institution Press, 1971.

Hacker, Barton C., and James M. Grimwood. *On the Shoulders of Titans: A History of Project Gemini*. NASA SP-4203, 1977; rep. ed. 2002.

Benson, Charles D., and William Barnaby Faherty. *Moonport: A History of Apollo Launch Facilities and Operations*. NASA SP-4204, 1978.

Brooks, Courtney G., James M. Grimwood, and Loyd S. Swenson, Jr. *Chariots for Apollo: A History of Manned Lunar Spacecraft*. NASA SP-4205, 1979.

Bilstein, Roger E. *Stages to Saturn: A Technological History of the Apollo/Saturn Launch Vehicles*. NASA SP-4206, 1980 and 1996.

No SP-4207.

Compton, W. David, and Charles D. Benson. *Living and Working in Space: A History of Skylab*. NASA SP-4208, 1983.

Ezell, Edward Clinton, and Linda Neuman Ezell. *The Partnership: A History of the Apollo-Soyuz Test Project*. NASA SP-4209, 1978.

Hall, R. Cargill. *Lunar Impact: A History of Project Ranger*. NASA SP-4210, 1977.

Newell, Homer E. *Beyond the Atmosphere: Early Years of Space Science*. NASA SP-4211, 1980.

Ezell, Edward Clinton, and Linda Neuman Ezell. *On Mars: Exploration of the Red Planet, 1958–1978*. NASA SP-4212, 1984.

Pitts, John A. *The Human Factor: Biomedicine in the Manned Space Program to 1980*. NASA SP-4213, 1985.

Compton, W. David. *Where No Man Has Gone Before: A History of Apollo Lunar Exploration Missions*. NASA SP-4214, 1989.

Naugle, John E. *First Among Equals: The Selection of NASA Space Science Experiments*. NASA SP-4215, 1991.

Wallace, Lane E. *Airborne Trailblazer: Two Decades with NASA Langley's 737 Flying Laboratory*. NASA SP-4216, 1994.

Butrica, Andrew J., ed. *Beyond the Ionosphere: Fifty Years of Satellite Communications*. NASA SP-4217, 1997.

Butrica, Andrew J. *To See the Unseen: A History of Planetary Radar Astronomy*. NASA SP-4218, 1996.

Mack, Pamela E., ed. *From Engineering Science to Big Science: The NACA and NASA Collier Trophy Research Project Winners*. NASA SP-4219, 1998.

Reed, R. Dale. *Wingless Flight: The Lifting Body Story*. NASA SP-4220, 1998.

Heppenheimer, T. A. *The Space Shuttle Decision: NASA's Search for a Reusable Space Vehicle*. NASA SP-4221, 1999.

Hunley, J. D., ed. *Toward Mach 2: The Douglas D-558 Program*. NASA SP-4222, 1999.

Swanson, Glen E., ed. *"Before This Decade Is Out…" Personal Reflections on the Apollo Program*. NASA SP-4223, 1999.

Tomayko, James E. *Computers Take Flight: A History of NASA's Pioneering Digital Fly-By-Wire Project*. NASA SP-4224, 2000.

Morgan, Clay. *Shuttle-Mir: The United States and Russia Share History's Highest Stage*. NASA SP-2001-4225, 2001.

Leary, William M. *"We Freeze to Please": A History of NASA's Icing Research Tunnel and the Quest for Safety*. NASA SP-2002-4226, 2002.

Mudgway, Douglas J. *Uplink-Downlink: A History of the Deep Space Network, 1957–1997.* NASA SP-2001-4227, 2001.

No SP-4228 or SP-4229.

Dawson, Virginia P., and Mark D. Bowles. *Taming Liquid Hydrogen: The Centaur Upper Stage Rocket, 1958–2002.* NASA SP-2004-4230, 2004.

Meltzer, Michael. *Mission to Jupiter: A History of the Galileo Project.* NASA SP-2007-4231, 2007.

Heppenheimer, T. A. *Facing the Heat Barrier: A History of Hypersonics.* NASA SP-2007-4232, 2007.

Tsiao, Sunny. *"Read You Loud and Clear!" The Story of NASA's Spaceflight Tracking and Data Network.* NASA SP-2007-4233, 2007.

Meltzer, Michael. *When Biospheres Collide: A History of NASA's Planetary Protection Programs.* NASA SP-2011-4234, 2011.

Center Histories, NASA SP-4300

Rosenthal, Alfred. *Venture into Space: Early Years of Goddard Space Flight Center.* NASA SP-4301, 1985.

Hartman, Edwin P. *Adventures in Research: A History of Ames Research Center, 1940–1965.* NASA SP-4302, 1970.

Hallion, Richard P. *On the Frontier: Flight Research at Dryden, 1946–1981.* NASA SP-4303, 1984.

Muenger, Elizabeth A. *Searching the Horizon: A History of Ames Research Center, 1940–1976.* NASA SP-4304, 1985.

Hansen, James R. *Engineer in Charge: A History of the Langley Aeronautical Laboratory, 1917–1958.* NASA SP-4305, 1987.

Dawson, Virginia P. *Engines and Innovation: Lewis Laboratory and American Propulsion Technology.* NASA SP-4306, 1991.

Dethloff, Henry C. *"Suddenly Tomorrow Came…": A History of the Johnson Space Center, 1957–1990.* NASA SP-4307, 1993.

Hansen, James R. *Spaceflight Revolution: NASA Langley Research Center from Sputnik to Apollo.* NASA SP-4308, 1995.

Wallace, Lane E. *Flights of Discovery: An Illustrated History of the Dryden Flight Research Center.* NASA SP-4309, 1996.

Herring, Mack R. *Way Station to Space: A History of the John C. Stennis Space Center.* NASA SP-4310, 1997.

Wallace, Harold D., Jr. *Wallops Station and the Creation of an American Space Program.* NASA SP-4311, 1997.

Wallace, Lane E. *Dreams, Hopes, Realities. NASA's Goddard Space Flight Center: The First Forty Years.* NASA SP-4312, 1999.

Dunar, Andrew J., and Stephen P. Waring. *Power to Explore: A History of Marshall Space Flight Center, 1960–1990.* NASA SP-4313, 1999.

Bugos, Glenn E. *Atmosphere of Freedom: Sixty Years at the NASA Ames Research Center.* NASA SP-2000-4314, 2000.

Bugos, Glenn E. *Atmosphere of Freedom: Seventy Years at the NASA Ames Research Center.* NASA SP-2010-4314, 2010. Revised version of NASA SP-2000-4314.

Bugos, Glenn E. *Atmosphere of Freedom: Seventy Five Years at the NASA Ames Research Center.* NASA SP-2014-4314, 2014. Revised version of NASA SP-2000-4314.

No SP-4315.

Schultz, James. *Crafting Flight: Aircraft Pioneers and the Contributions of the Men and Women of NASA Langley Research Center.* NASA SP-2003-4316, 2003.

Bowles, Mark D. *Science in Flux: NASA's Nuclear Program at Plum Brook Station, 1955–2005.* NASA SP-2006-4317, 2006.

Wallace, Lane E. *Flights of Discovery: An Illustrated History of the Dryden Flight Research Center.* NASA SP-2007-4318, 2007. Revised version of NASA SP-4309.

Arrighi, Robert S. *Revolutionary Atmosphere: The Story of the Altitude Wind Tunnel and the Space Power Chambers.* NASA SP-2010-4319, 2010.

General Histories, NASA SP-4400

Corliss, William R. *NASA Sounding Rockets, 1958–1968: A Historical Summary.* NASA SP-4401, 1971.

Wells, Helen T., Susan H. Whiteley, and Carrie Karegeannes. *Origins of NASA Names.* NASA SP-4402, 1976.

Anderson, Frank W., Jr. *Orders of Magnitude: A History of NACA and NASA, 1915–1980.* NASA SP-4403, 1981.

Sloop, John L. *Liquid Hydrogen as a Propulsion Fuel, 1945–1959.* NASA SP-4404, 1978.

Roland, Alex. *A Spacefaring People: Perspectives on Early Spaceflight.* NASA SP-4405, 1985.

Bilstein, Roger E. *Orders of Magnitude: A History of the NACA and NASA, 1915–1990.* NASA SP-4406, 1989.

Logsdon, John M., ed., with Linda J. Lear, Jannelle Warren Findley, Ray A. Williamson, and Dwayne A. Day. *Exploring the Unknown: Selected Documents in the History of the U.S. Civil Space Program, Volume I: Organizing for Exploration.* NASA SP-4407, 1995.

Logsdon, John M., ed., with Dwayne A. Day and Roger D. Launius. *Exploring the Unknown: Selected Documents in the History of the U.S. Civil Space Program, Volume II: External Relationships.* NASA SP-4407, 1996.

Logsdon, John M., ed., with Roger D. Launius, David H. Onkst, and Stephen J. Garber. *Exploring the Unknown: Selected Documents in the History of the U.S. Civil Space Program, Volume III: Using Space.* NASA SP-4407, 1998.

Logsdon, John M., ed., with Ray A. Williamson, Roger D. Launius, Russell J. Acker, Stephen J. Garber, and Jonathan L. Friedman. *Exploring the Unknown: Selected Documents in the History of the U.S. Civil Space Program, Volume IV: Accessing Space.* NASA SP-4407, 1999.

Logsdon, John M., ed., with Amy Paige Snyder, Roger D. Launius, Stephen J. Garber, and Regan Anne Newport. *Exploring the Unknown: Selected Documents in the History of the U.S. Civil Space Program, Volume V: Exploring the Cosmos.* NASA SP-2001-4407, 2001.

Logsdon, John M., ed., with Stephen J. Garber, Roger D. Launius, and Ray A. Williamson. *Exploring the Unknown: Selected Documents in the History of the U.S. Civil Space Program, Volume VI: Space and Earth Science.* NASA SP-2004-4407, 2004.

Logsdon, John M., ed., with Roger D. Launius. *Exploring the Unknown: Selected Documents in the History of the U.S. Civil Space Program, Volume VII: Human Spaceflight: Projects Mercury, Gemini, and Apollo.* NASA SP-2008-4407, 2008.

Siddiqi, Asif A., *Challenge to Apollo: The Soviet Union and the Space Race, 1945–1974.* NASA SP-2000-4408, 2000.

Hansen, James R., ed. *The Wind and Beyond: Journey into the History of Aerodynamics in America, Volume 1: The Ascent of the Airplane.* NASA SP-2003-4409, 2003.

Hansen, James R., ed. *The Wind and Beyond: Journey into the History of Aerodynamics in America, Volume 2: Reinventing the Airplane.* NASA SP-2007-4409, 2007.

Hogan, Thor. *Mars Wars: The Rise and Fall of the Space Exploration Initiative.* NASA SP-2007-4410, 2007.

Vakoch, Douglas A., ed. *Psychology of Space Exploration: Contemporary Research in Historical Perspective.* NASA SP-2011-4411, 2011.

Ferguson, Robert G. *NASA's First A: Aeronautics from 1958 to 2008.* NASA SP-2012-4412, 2013.

Vakoch, Douglas A., ed. *Archaeology, Anthropology, and Interstellar Communication.* NASA SP-2013-4413, 2014.

Asner, Glen R., and Stephen J. Garber. *Origins of 21st-Century Space Travel: A History of NASA's Decadal Planning Team and the Vision for Space Exploration, 1999–2004.* NASA SP-2019-4415, 2019.

Monographs in Aerospace History, NASA SP-4500

Launius, Roger D., and Aaron K. Gillette, comps. *Toward a History of the Space Shuttle: An Annotated Bibliography.* Monographs in Aerospace History, No. 1, 1992.

Launius, Roger D., and J. D. Hunley, comps. *An Annotated Bibliography of the Apollo Program.* Monographs in Aerospace History, No. 2, 1994.

Launius, Roger D. *Apollo: A Retrospective Analysis.* Monographs in Aerospace History, No. 3, 1994.

Hansen, James R. *Enchanted Rendezvous: John C. Houbolt and the Genesis of the Lunar-Orbit Rendezvous Concept.* Monographs in Aerospace History, No. 4, 1995.

Gorn, Michael H. *Hugh L. Dryden's Career in Aviation and Space.* Monographs in Aerospace History, No. 5, 1996.

Powers, Sheryll Goecke. *Women in Flight Research at NASA Dryden Flight Research Center from 1946 to 1995.* Monographs in Aerospace History, No. 6, 1997.

Portree, David S. F., and Robert C. Trevino. *Walking to Olympus: An EVA Chronology.* Monographs in Aerospace History, No. 7, 1997.

Logsdon, John M., moderator. *Legislative Origins of the National Aeronautics and Space Act of 1958: Proceedings of an Oral History Workshop.* Monographs in Aerospace History, No. 8, 1998.

Rumerman, Judy A., comp. *U.S. Human Spaceflight: A Record of Achievement, 1961–1998.* Monographs in Aerospace History, No. 9, 1998.

Portree, David S. F. *NASA's Origins and the Dawn of the Space Age.* Monographs in Aerospace History, No. 10, 1998.

Logsdon, John M. *Together in Orbit: The Origins of International Cooperation in the Space Station.* Monographs in Aerospace History, No. 11, 1998.

Phillips, W. Hewitt. *Journey in Aeronautical Research: A Career at NASA Langley Research Center.* Monographs in Aerospace History, No. 12, 1998.

Braslow, Albert L. *A History of Suction-Type Laminar-Flow Control with Emphasis on*

Flight Research. Monographs in Aerospace History, No. 13, 1999.

Logsdon, John M., moderator. *Managing the Moon Program: Lessons Learned from Apollo.* Monographs in Aerospace History, No. 14, 1999.

Perminov, V. G. *The Difficult Road to Mars: A Brief History of Mars Exploration in the Soviet Union.* Monographs in Aerospace History, No. 15, 1999.

Tucker, Tom. *Touchdown: The Development of Propulsion Controlled Aircraft at NASA Dryden.* Monographs in Aerospace History, No. 16, 1999.

Maisel, Martin, Demo J. Giulanetti, and Daniel C. Dugan. *The History of the XV-15 Tilt Rotor Research Aircraft: From Concept to Flight.* Monographs in Aerospace History, No. 17, 2000. NASA SP-2000-4517.

Jenkins, Dennis R. *Hypersonics Before the Shuttle: A Concise History of the X-15 Research Airplane.* Monographs in Aerospace History, No. 18, 2000. NASA SP-2000-4518.

Chambers, Joseph R. *Partners in Freedom: Contributions of the Langley Research Center to U.S. Military Aircraft of the 1990s.* Monographs in Aerospace History, No. 19, 2000. NASA SP-2000-4519.

Waltman, Gene L. *Black Magic and Gremlins: Analog Flight Simulations at NASA's Flight Research Center.* Monographs in Aerospace History, No. 20, 2000. NASA SP-2000-4520.

Portree, David S. F. *Humans to Mars: Fifty Years of Mission Planning, 1950–2000.* Monographs in Aerospace History, No. 21, 2001. NASA SP-2001-4521.

Thompson, Milton O., with J. D. Hunley. *Flight Research: Problems Encountered and What They Should Teach Us.* Monographs in Aerospace History, No. 22, 2001. NASA SP-2001-4522.

Tucker, Tom. *The Eclipse Project.* Monographs in Aerospace History, No. 23, 2001. NASA SP-2001-4523.

Siddiqi, Asif A. *Deep Space Chronicle: A Chronology of Deep Space and Planetary Probes, 1958–2000.* Monographs in Aerospace History, No. 24, 2002. NASA SP-2002-4524.

Merlin, Peter W. *Mach 3+: NASA/USAF YF-12 Flight Research, 1969–1979.* Monographs in Aerospace History, No. 25, 2001. NASA SP-2001-4525.

Anderson, Seth B. *Memoirs of an Aeronautical Engineer: Flight Tests at Ames Research Center: 1940–1970.* Monographs in Aerospace History, No. 26, 2002. NASA SP-2002-4526.

Renstrom, Arthur G. *Wilbur and Orville Wright: A Bibliography Commemorating the One-Hundredth Anniversary of the First Powered Flight on December 17, 1903.* Monographs in Aerospace History, No. 27, 2002. NASA SP-2002-4527.

No monograph 28.

Chambers, Joseph R. *Concept to Reality: Contributions of the NASA Langley Research Center to U.S. Civil Aircraft of the 1990s.* Monographs in Aerospace History, No. 29, 2003. NASA SP-2003-4529.

Peebles, Curtis, ed. *The Spoken Word: Recollections of Dryden History, The Early Years.* Monographs in Aerospace History, No. 30, 2003. NASA SP-2003-4530.

Jenkins, Dennis R., Tony Landis, and Jay Miller. *American X-Vehicles: An Inventory—X-1 to X-50.* Monographs in Aerospace History, No. 31, 2003. NASA SP-2003-4531.

Renstrom, Arthur G. *Wilbur and Orville Wright: A Chronology Commemorating the One-Hundredth Anniversary of the First Powered Flight on December 17, 1903.* Monographs in Aerospace History, No. 32, 2003. NASA SP-2003-4532.

Bowles, Mark D., and Robert S. Arrighi. *NASA's Nuclear Frontier: The Plum Brook Research Reactor.* Monographs in Aerospace History, No. 33, 2004. NASA SP-2004-4533.

Wallace, Lane, and Christian Gelzer. *Nose Up: High Angle-of-Attack and Thrust Vectoring Research at NASA Dryden, 1979–2001.* Monographs in Aerospace History, No. 34, 2009. NASA SP-2009-4534.

Matranga, Gene J., C. Wayne Ottinger, Calvin R. Jarvis, and D. Christian Gelzer. *Unconventional, Contrary, and Ugly: The Lunar Landing Research Vehicle.* Monographs in Aerospace History, No. 35, 2006. NASA SP-2004-4535.

McCurdy, Howard E. *Low-Cost Innovation in Spaceflight: The History of the Near Earth Asteroid Rendezvous (NEAR) Mission.* Monographs in Aerospace History, No. 36, 2005. NASA SP-2005-4536.

Seamans, Robert C., Jr. *Project Apollo: The Tough Decisions.* Monographs in Aerospace History, No. 37, 2005. NASA SP-2005-4537.

Lambright, W. Henry. *NASA and the Environment: The Case of Ozone Depletion.* Monographs in Aerospace History, No. 38, 2005. NASA SP-2005-4538.

Chambers, Joseph R. *Innovation in Flight: Research of the NASA Langley Research Center on Revolutionary Advanced Concepts for Aeronautics.* Monographs in Aerospace History, No. 39, 2005. NASA SP-2005-4539.

Phillips, W. Hewitt. *Journey into Space Research: Continuation of a Career at NASA Langley Research Center.* Monographs in Aerospace History, No. 40, 2005. NASA SP-2005-4540.

Rumerman, Judy A., Chris Gamble, and Gabriel Okolski, comps. *U.S. Human Spaceflight: A Record of Achievement, 1961–2006.* Monographs in Aerospace History, No. 41, 2007. NASA SP-2007-4541.

Peebles, Curtis. *The Spoken Word: Recollections of Dryden History Beyond the Sky.* Monographs in Aerospace History, No. 42, 2011. NASA SP-2011-4542.

Dick, Steven J., Stephen J. Garber, and Jane H. Odom. *Research in NASA History.* Monographs in Aerospace History, No. 43, 2009. NASA SP-2009-4543.

Merlin, Peter W. *Ikhana: Unmanned Aircraft System Western States Fire Missions.* Monographs in Aerospace History, No. 44, 2009. NASA SP-2009-4544.

Fisher, Steven C., and Shamim A. Rahman. *Remembering the Giants: Apollo Rocket Propulsion Development.* Monographs in Aerospace History, No. 45, 2009. NASA SP-2009-4545.

Gelzer, Christian. *Fairing Well: From Shoebox to Bat Truck and Beyond, Aerodynamic Truck Research at NASA's Dryden Flight Research Center.* Monographs in Aerospace History, No. 46, 2011. NASA SP-2011-4546.

Arrighi, Robert. *Pursuit of Power: NASA's Propulsion Systems Laboratory No. 1 and 2.* Monographs in Aerospace History, No. 48, 2012. NASA SP-2012-4548.

Renee M. Rottner. *Making the Invisible Visible: A History of the Spitzer Infrared Telescope Facility (1971–2003).* Monographs in Aerospace History, No. 47, 2017. NASA SP-2017-4547.

Goodrich, Malinda K., Alice R. Buchalter, and Patrick M. Miller, comps. *Toward a History of the Space Shuttle: An Annotated Bibliography, Part 2 (1992–2011)*. Monographs in Aerospace History, No. 49, 2012. NASA SP-2012-4549.

Ta, Julie B., and Robert C. Treviño. *Walking to Olympus: An EVA Chronology, 1997–2011*, Vol. 2. Monographs in Aerospace History, No. 50, 2016. NASA SP-2016-4550.

Gelzer, Christian. *The Spoken Word III: Recollections of Dryden History; The Shuttle Years*. Monographs in Aerospace History, No. 52, 2013. NASA SP-2013-4552.

Ross, James C. *NASA Photo One*. Monographs in Aerospace History, No. 53, 2013. NASA SP-2013-4553.

Launius, Roger D. *Historical Analogs for the Stimulation of Space Commerce*. Monographs in Aerospace History, No. 54, 2014. NASA SP-2014-4554.

Buchalter, Alice R., and Patrick M. Miller, comps. *The National Advisory Committee for Aeronautics: An Annotated Bibliography*. Monographs in Aerospace History, No. 55, 2014. NASA SP-2014-4555.

Chambers, Joseph R., and Mark A. Chambers. *Emblems of Exploration: Logos of the NACA and NASA*. Monographs in Aerospace History, No. 56, 2015. NASA SP-2015-4556.

Alexander, Joseph K. *Science Advice to NASA: Conflict, Consensus, Partnership, Leadership*. Monographs in Aerospace History, No. 57, 2017. NASA SP-2017-4557.

Electronic Media, NASA SP-4600

Remembering Apollo 11: The 30th Anniversary Data Archive CD-ROM. NASA SP-4601, 1999.

Remembering Apollo 11: The 35th Anniversary Data Archive CD-ROM. NASA SP-2004-4601, 2004. This is an update of the 1999 edition.

The Mission Transcript Collection: U.S. Human Spaceflight Missions from Mercury Redstone 3 to Apollo 17. NASA SP-2000-4602, 2001.

Shuttle-Mir: The United States and Russia Share History's Highest Stage. NASA SP-2001-4603, 2002.

U.S. Centennial of Flight Commission Presents Born of Dreams—Inspired by Freedom. NASA SP-2004-4604, 2004.

Of Ashes and Atoms: A Documentary on the NASA Plum Brook Reactor Facility. NASA SP-2005-4605, 2005.

Taming Liquid Hydrogen: The Centaur Upper Stage Rocket Interactive CD-ROM. NASA SP-2004-4606, 2004.

Fueling Space Exploration: The History of NASA's Rocket Engine Test Facility DVD. NASA SP-2005-4607, 2005.

Altitude Wind Tunnel at NASA Glenn Research Center: An Interactive History CD-ROM. NASA SP-2008-4608, 2008.

A Tunnel Through Time: The History of NASA's Altitude Wind Tunnel. NASA SP-2010-4609, 2010.

Conference Proceedings, NASA SP-4700

Dick, Steven J., and Keith Cowing, eds. *Risk and Exploration: Earth, Sea and the Stars*. NASA SP-2005-4701, 2005.

Dick, Steven J., and Roger D. Launius. *Critical Issues in the History of Spaceflight*. NASA SP-2006-4702, 2006.

Dick, Steven J., ed. *Remembering the Space Age: Proceedings of the 50th Anniversary Conference*. NASA SP-2008-4703, 2008.

Dick, Steven J., ed. *NASA's First 50 Years: Historical Perspectives*. NASA SP-2010-4704, 2010.

Societal Impact, NASA SP-4800

Dick, Steven J., and Roger D. Launius. *Societal Impact of Spaceflight*. NASA SP-2007-4801, 2007.

Dick, Steven J., and Mark L. Lupisella. *Cosmos and Culture: Cultural Evolution in a Cosmic Context*. NASA SP-2009-4802, 2009.

Dick, Steven J. *Historical Studies in the Societal Impact of Spaceflight*. NASA SP-2015-4803, 2015.

INDEX